and the answer is a pineapple

The King of fruit in folklore, fabric and food

Claudia Hyles

MILNER PUBLISHING

To Geoffrey. Jamie and Emily

First published in 1998 by Sally Milner Publishing Pty Ltd

1423 Burra Road
Burra Creek 2620
New South Wales
Australia

Designed by Amanda Knobel, ANU Graphics, Canberra

Printed and bound in China

National Library of Australia Cataloguing in Publication data:

Hyles, Claudia.

And the answer is a pineapple.

Bibliography.

ISBN 1 86351 221 7.

1. Pineapple. 2. Cookery (pineapples). I. Title 641.34774

Locations in pursuit of the Pineapple

Preface

Kongthong Nanthavongdouangsy, a famous weaver from Laos, sent me a Lao riddle which asks 'What does it mean when the Red Cock lies down by the edge of the fence?' 'The answer is a pineapple.' I'm afraid I can't quite understand this, but it's odd that the answer is the title of my book.

The title didn't come from Laos, but rather from my dear friend, Richard Miller, in Canberra, Australia. For years his mother, Jo, used to broadcast 'Turf Talk' from 3UZ in Melbourne, every Saturday morning with Bert Bryant and Jack Ayling. The programme was so entertaining and funny, as well as accurate about picking the day's winners, that a spin-off went to air for a while in the mid-'70s. Before the era of talk-back radio, it was based on letters from listeners, some serious, but most crazy. The name came from Bert Bryant's stock reply to any stumping question — 'And the Answer is a Pineapple'. Great name for a book.

The one thing I always wanted to do, from the time I was a child, was to write a book. As I grew older, the charms of Ethel Turner and Enid Blyton ebbed and I imagined writing an earth-shattering novel — Booker prize material, not a series of scribbles about a tropical fruit.

Pineapples were always in the home in which I grew up — first in a cut-glass bowl with the apples and oranges and bananas, and then, peeled and chopped and placed in the 'fridge in a Bushell's coffee jar of thick glass with a screw-top lid — I think with a few spoonfuls of sugar. My mother was and is an imaginative and adventurous cook and we didn't sit down to meat and three vegetables night after night. Pineapple, I remember, featured in her delicious version of Chow Mein. She is also a keen gardener and that is another inherited interest.

The idea of writing about pineapples first dropped into my head while living in Laos. We were there for three and a half years and for me it was a very creative time. I learned about the superb textiles, dabbled in various art forms, travelled to exciting places and did things in my own right. During this time, I worked as Cultural Relations Officer in the Australian Embassy in Bangkok, as a volunteer for the Australian National Library in the National Library of Laos and with the aid of a friend, I founded the Women's International Club in Vientiane in 1990. It is still going strong and has done a tremendous amount to help both needy Lao women and children and expatriate women who might feel cut adrift at suddenly finding themselves in a strange, new environment.

My adventures abroad started in 1968 when I went alone to India. I had fallen in love with it when I was about ten. I read, with a certain degree of perseverance, as the chapters were scattered through the 1923 issue of *Chatterbox*, a story called 'Louisa' by Mrs Hobart-Hampden. It was the adventures of a little memsahib and I thought it most exciting. In my final year of school, I studied the struggle for India's Independence in Modern History. Names and events like Amritsar, The Salt March and Chauri Chaura stuck in my head and I met a member of my extended family who had met Jawarharlal Nehru.

Then I read *A Passage to India* at University and thought, 'That's it, I'll go'.

On that trip, I became aware of traditional craft and above all, handloom textiles in all their beauty. I discovered I could become a chameleon by wearing a sari or shalwar kamiz and, far from getting thin or sick, I adored the food and put on weight. I have been to India ten times since.

In the early '70s I married an Australian farmer and we thought we would be staying on the land for the rest of our lives. The rural recession struck in the mid-'70s however, and our future changed. After years of study, my husband worked for the Ministry of Finance in the Government of Papua New Guinea, and we had a happy two years in Port Moresby. With the children, we travelled within Papua New Guinea and to Hong Kong and the Philippines.

In Port Moresby, I worked as a volunteer at the Art Gallery and Museum, doing all sorts of interesting things — mapping and editorial work for Museum publications—and came face to face with traditional tribal art for the first time. This work set me on a wonderful path and I have worked in museums and art galleries, absorbing history and culture in the easiest way, in most countries in which I've lived. The year's study at the National Gallery of Australia in 1983 in order to become a guide, was the most marvellous and stimulating time and I think it has set me up for life.

A year in Great Britain was followed eventually by two years in Trinidad and Tobago, where Geoffrey worked for the Food and Agriculture Organisation of the United Nations. While there, we travelled within the Caribbean and to parts of the United States of America and Central and South America. I guess the pineapple rose in my sub-conscious there, especially as they grew in our garden.

In Pakistan and Laos, Geoffrey worked for the United Nations Drug Control Programme on opium poppy eradication projects, and then in Somalia and Vietnam for the United Nations Development Programme.

In between all the stints abroad we would come back to our farm near Canberra. We would usually have to resurrect the garden, Geoffrey would be reunited with his cattle, and I would return to the Gallery and catering and whatever happened next.

Wherever I've lived, I've cooked and read. In some temporary homes I have been able to play tennis or mah jong, and sew or swim. In others, to sing and garden, to write, and even run. I learnt the piano for years as a child, but I didn't always have one to play while living abroad. There are often things one misses and life in the tropics isn't just a gin and tonic under a palm tree.

It was in Nairobi where we had to live during the Somalia interlude that I decided to really knuckle down and start writing. We went to a party and somebody asked me what I did. Instead of mewing about this and that, I said I was writing a book. She happened to be a publisher and asked about the subject. 'Pineapples', I answered. She laughed and said, 'It's a great thing people are writing these whacky books!' So here it is — the whacky book.

Recipes

When I asked my Ghanaian friend if pineapples had any special significance in her country she was very dismissive indeed. 'No', she said. 'In the Twi and Ga languages the name for the fruit and the plant is "the European palm". No there's nothing interesting about the pineapple — they're good to eat though.'

When I asked her, I was living in Kenya and we were in an Ethiopian Airways plane on our way to Accra, the capital of Ghana. I spent two weeks in her country and loved it. What's more I found out some fascinating things about the pineapple. How wrong my friend was about the significance of this fruit.

Nobody knows just who brought the pineapple to Ghana, but contact with Europeans, who undoubtedly introduced the fruit, goes back hundreds of years to the days when the country was known as the Gold Coast. For centuries the trade routes from West Africa to the rest of the world were northward-looking — overland routes rather than sea. However the voyages of exploration, in particular of the Portuguese and Spanish in the fifteenth century, started to change this, and later the demand for African slaves for the new colonies in the Americas saw the establishment of fortified trading posts on the Atlantic Coast.

There are about 50 of these forts or castles along the sandy palm-fringed beaches. I was able to visit three of them — Cape Coast, Elmina and Christianborg in Accra.

Elmina was the first to have been built — by a Portuguese, Don Diego de Azambuja, between 1471 and 1482. It is white and solid and situated at the western end of a wide, sweeping bay. One's first view of it is very picturesque, framed by coconut palms and the colourful flags of fishermen's boats. The Atlantic is usually a rather sullen, grey colour but the day of my visit to Elmina it was a perfect, limpid, blue-green.

It is a beautiful setting but such a sad place. The Dutch who ousted the Portuguese from the castle changed the chapel into a slave market. Another change was made to the doors from which cargo was transferred to the ships which would travel halfway around the world. The doors were made narrower so the human cargo could not easily escape.

The two young African Americans who were among the small group of visitors were very moved by the atmosphere. I kept thinking of my friends in Trinidad who are of African descent, some of whom still have names that come from places such as Ghana. I felt as though I was in Trinidad so often in those two weeks, perhaps because I had learned in Trinidad the little I knew about Ghana — from my husband's then boss, and his wife, from Ralph and Wilma Hoyte who had worked there as doctors, and from Emmanuel Asibey, the Director of Ghana's Wildlife Department, working temporarily in Port-of-Spain, who came to our dinners in his kente cloth robes.

My first serious pineapple sleuthing was at the National Museum in Accra, where I discovered carved wooden umbrella tops, or *kyinie akyi*, which surmount the ceremonial parasols used to shade Asante chiefs at important gatherings. Umbrella tops help to distinguish a paramount chief from his sub-chiefs at a public gathering and signal his presence or arrival. Usually wrapped in gold or silver leaf, these tops represent proverbial sayings, historical events or the qualities which rulers are said to possess. A particular occasion may call for a particular umbrella which has a finial appropriate to the event.

Asante umbrellas are still perhaps the most striking part of the royal Ghanaian regalia. Their coverings are spectacular and the most popular colour was red, patterned and plain, velvet and silk. The umbrella of course had a sound physical use—to keep the King cool and shaded—but it also created a zone of calm and peace. A well-known saying is that 'God should not see the crown of the King's head', so the umbrella acts as a barrier between the heavens and the ruler as his sandals and footstool prevent his physical contact with the earth beneath him.

A top carved as a pineapple with little shooting fruits around the main fruit has two meanings: that a good family will never be uprooted; and that there is no good turn without much effort. Another pineapple top that I saw in the Kumasi Museum had the symbolic meaning of 'God's time is the best'.

Kumasi is the old capital of the Asante kingdom. It is situated 200 kilometres or so from Accra, in an area of dense tropical rainforest. I travelled there by crowded public bus on a seat which folded down into the central aisle. The journey took about four hours and the views of the forest were wonderful with huge and majestic trees soaring up from a chaotic profusion of oil palms, cocoa trees, cassava and low-growing crops of vegetables. We passed through towns and villages where the walls of the iron-roofed houses were stained red from the lateritic earth. There was something exciting in the atmosphere and Kumasi has a proud wildness.

The first account of the court of the King of the Asante by an Englishman, was written in 1817 by T. E. Bowdich, who was quite dazzled by its splendour. After this time, British contact was punctuated by wars. Kumasi was sacked in the Fourth Ashanti War in 1874. The ruler was exiled to the Seychelles in 1896 and not restored to the throne, or as it is in Kumasi, the stool, until 1935.

The present Asantehene, Opoku Ware II is a lawyer by profession. He had just returned after a sojourn in England when I arrived in Kumasi. George, a very nice taxi driver and owner of the 'Negative Ideas' taxi, suggested that I go to the Palace office and see if I could have a ticket for the durbar the next day, when the Asantehene would hold a public audience.

While I waited clerks asked me if I would like to buy a commemorative t-shirt or hat. I didn't know what they were commemorating but found to my delight it was the silver jubilee of the Asantehene. It so happened that it was also the silver jubilee of the year my husband and I were wed so I joyfully bought three t-shirts, one for each of our children and one for their father.

The Asantehene's face appears on the front and 'Silver Jubilee 1970-1995' on the back.

Even more exciting was being able to witness the durbar. There was the King on his throne, adorned with gold, shaded by a green velvet umbrella surmounted by a golden finial, his feet, encased in gold-adorned sandals, resting on a stool. An attendant fanned him occasionally and vigorously with a large palm-leaf fan while raucous peacocks squawked from the roof of the modern palace building.

In a deep semi-circle around the Asantehene sat men who were chiefs. Bowdich described the robes of the chiefs he saw as 'Ashantee cloths of extravagant price ... of incredible size and weight thrown over the shoulder exactly like a Roman toga'. Still it is today. The men looked like a collection of Roman senators in mostly dark colours — black and rich, dark aubergine. There was a sprinkling of lighter colours and here and there a gold watch gleamed on dark skin. Apart from the gold watches and a microphone one could have just stepped into one of those old photographs taken at the Asantehene's court 100 years before.

No-one addressed the King directly and, traditionally, a linguist or spokesman repeats, in diplomatic and often very witty language, the remarks of all involved in a meeting or discussion. He normally carries a staff or *kyame poma* although I couldn't see it this day. These staffs have become almost as elaborate as the umbrella finials. They are made in the same way — of carved wood, covered in gold leaf, or sometimes silver. Animals, humans and all kinds of objects, including the pineapple, have been used as images. Like the umbrella tops, the staff finial is connected to a proverb or aphorism.

The linking of a proverb with a visual image is an important tradition, not only in Asante culture, but all over Ghana. Where there is no written language, and the codes of a society are handed on by oral tradition, the richness of this enormous corpus of aphorisms, adages and *bon mots* reinforces the collective memory. In modern Ghana the tradition is still important.

This was the first large gathering of traditionally dressed people that I saw, but certainly not the last. A convocation at the University in Accra was the most splendid congregation and out in the countryside, at very regular intervals, I came across funerals. 'Funerals will be the death of Ghana', my friends said to me. They certainly are big events and cost the family of the deceased person a great deal of money.

In Kumasi I stayed in a hotel called Sir Max which was very comfortable. In the evenings I entertained myself, and I hope not anyone else, by singing. While I was away in Ghana, I was missing two important rehearsals of the Nairobi Music Society Choir. We were singing 'Carmina Burana' at two performances only two days after my return, so I was anxious not to make any dreadful blunders in the second sopranos. I had a little tape recorder and each night in Kumasi and in Accra I would warble away at those wonderful songs.

Near Kumasi is a village called Ntonsu which is a centre for the weaving and printing of adinkra cloth. Like other towns and villages in this area it had rather a wild west atmosphere. Bowdich collected a piece of adinkra in 1817 which was stamped with the same symbols used today. Like the umbrella finials the designs have a corresponding proverb or aphorism — everything is full of hidden meanings in Ghana. The designs are applied to the cloth with stamps cut from a piece of calabash or gourd. These are attached to two or three short sticks bound together as a handle. Then they are dipped into a tarry mixture or 'medicine' made from bark with an iron mordant.

A funeral cloth from Ghana, West Africa

I had come to Ntonsu to see the adinkra but found that dyeing of the other machine-woven cloth to shades appropriate for funeral wear was just as important in the village. The vats of bark medicine for the adinkra were boiling away next to the funeral dye vats, and this dye, which varied from rusty black through all shades of deep aubergine, was made from roots.

Back in Accra, at the Makola Market, I discovered, however, that not everyone wears old robes redyed for funerals. The Market is always thronged with people. It was raining when I visited and the surrounding streets were muddy but vendors squatted down on the kerbs, selling from huge enamel basins the reddest tomatoes, okra, smoked fish and enormous snails. Many women wore at rakish angles great plaited palm-leaf hats like cart-wheels. The footpaths were crowded and the giant enamel basins, normally carried on the head, were, when empty, tipped

Buying cloth in the Makola Market in Accra, Ghana. Red and black cloth is worn for funerals and there are many designs, including the pineapple

sideways and pointing to the skies in order not to decapitate a passerby. Further into the market were pillars of these vessels — imported from China — placed one on top of each other, lip to lip, like colossal balusters.

Even further in was the cloth market. Ghanaian women have wonderful style and flair — it's not just the Roman toga-clad men who look so stunning. Both men and women wear the same kind of cloth, known locally as wax-cloth. For a man one buys 12 yards and for a woman six. From this six — and it is 120 cm (approximately 48 inches) wide — two is for a blouse, two for a skirt and two for a useful piece which can be a head wrapper, a waist wrapper, a baby sling or what you will.

The wax-cloths have a fascinating history. They were introduced to the African market by the Dutch and it is very plain to see from whence they sprang. Java, of course. Some of the patterns are straight from Javanese batiks. Some even have what purports to be Arabic calligraphy, but it is impossible to read it — it just looks like Arabic. The best quality cloth and the most expensive is still Dutch. The next is British, the next is the locally made Ghanaian cloth, and the cheapest, which is starting to flood into the market, is from China, just like the enamel basins.

I bought Ghanaian cloth, which is produced in two factories at the towns of Akosombo and Tema. While the Javanese heritage is apparent in these cloths, and one can see modified parang and other typical Indonesian designs, there is an overwhelming African flavour, which is exuberant and fun. All the colours of the rainbow are available but I concentrated on blue and white cloths, which in a group look rather like the Pekalongan Java batiks favoured by the Peranakan or Straits Chinese.

I was utterly hooked after my first morning at the markets and my first purchase was not blue and white, but rather red and black — a funeral cloth. Many graveyards are surrounded by shrubs of the croton family in colours of dark green, dark red and brown so this red and black colour choice for funerals must be related. The reds are wonderfully bright and the patterns certainly are not sombre in subject. Perhaps they represent the joys and pastimes of the deceased — or maybe the mourner — but I don't think so. They are so varied. There are snails, guinea-fowls, blackboards with the ABC, crossed tennis racquets with balls, and yes, of course, the pineapple.

I interviewed a number of vendors, some of whom sold only blue and white cloth, and tried to discover the names and meanings of all the pieces I bought. Motifs of food figured strongly. There was one showing ears of corn which was called 'Mother Sweet' because it is the mother in the family who provides the daily bread. One showing two peacocks with beaks touching was called 'When two lips come together there is no dispute'. There was 'Have Mercy on Me', 'It pains you', 'Hosiana', which was a paisley design, and 'Hallelujah, Amen, Praise the Lord'. One covered with stylised handprints was 'The back of the hand is sweeter than the palm', meaning, I was told, the grass is greener on the other side of the fence, but also surely it means that the palm administers the slap. My favourite was a design of circular shapes, larger at the hem and growing smaller as they advanced to the top — little oyster-like shapes really with a zig-zag line across the middle. I couldn't understand the vendor when she told me the name of this one so I asked her to explain and she said, 'In the 1960s we had an ointment which was very popular and the circles resemble the box'. The name of the design was 'Thermogen', the ointment being a deep heat rub.

In Ghana there are eight main languages and they are all full of aphorisms, proverbs and mottos. Ghanaians seem to be natural orators and they have a great sense of humour. They are also a very religious people. Some of the names of small businesses and institutions strike the visitor as very amusing. I saw the 'Wonderful God Grinding Spot', 'God the Creator Blockworks', 'Salon de Peace and Love', 'God Almighty Art Works', and 'The Sacred Cherubim and Seraphim Church'. I travelled, as I mentioned earlier, in a Kumasi taxi called 'Negative Ideas' for two days with no mishap, but I did not try 'Refreshing Pee Cola'!

If I had, perhaps I could have treated any bad symptoms with one of the pineapple remedies I learned of in Koforidua, a town about 50 km (30 miles) from Accra. I stayed there with Comfort and Hackman Owusu-Agyemang in a house surrounded by a garden full of vegetables and exotic fruits. There were custard apple trees and the strange ackee which I had eaten in the West Indies and which Harry Belafonte immortalised for everybody in his 'Jamaican Farewell'. Ackee rice is impossible to describe adequately — it has to be eaten.

Hack had been the Resident Representative of the Food and Agricultural Organisation in Trinidad when we were there and had been my husband's kind and considerate boss. I had a very enlightening weekend with him and his wife, learning a great deal about Ghanaian politics in opposition, among other things. But it was Comfort who told me of uses for the pineapple. As we strolled around the luxuriant garden, she explained that a treatment for jaundice is to drink the liquid obtained from boiling up pineapple skins in water. Another herbal medicine is made by boiling up the central part of the pineapple leaves when the plant is young.

I have since heard many other medicinal uses of the pineapple. An Indian friend says pineapple is good for arthritis, and my daughter swears that eating the fruit is a cure for hay fever. In Thai herbal medicine, juice extracted from the leaves can be taken as a taenifuge, which is an agent which expels tapeworm. The young fruit also acts to expel tapeworm in children, it is a diuretic and is said to be a cure for gonorrhoea. It also helps to breaks down mucus, and the juice is said to be a cure for hiccoughs. So, all in all, pineapples are good for you.

On our trip back to Accra from Koforidua, Comfort bought superb pineapples near the Aburi Botanical Gardens on Akwapim Ridge. Then as we neared Accra we really could have bought anything we liked while remaining seated in our comfortable car: mothballs, a puppy, handkerchiefs pinned around the edge of black umbrellas, popcorn, a hat, a shirt. All were available from vendors on the side of the road.

In the Kumasi Museum I bought a little cookbook called *Ghanaian Favourite Dishes — Recipes that are loved best in many Ghanaian homes*, written by Alice Dede. I have lots of cookbooks from countries such as Ghana where there are instructions for tending the fire and household hints for things that just don't exist in Australian homes. One of my Indian ones firmly advises that a good housewife must see that she has a good supply of pyjama strings in her almirah or cupboard. A Papua New Guinean one, written by a nun, has a great recipe for cooking a crocodile. The absolute worst is a compilation by a group of vegetarian American missionaries in Kenya who manage to murder all kinds of good food in a book called *Missionary Manna*.

The Ghanaian book, however, is good although one needs to have a penchant for the huge amounts of carbohydrates consumed there. Steamed cassava dough and kenkey, which is steamed corn dough, are popular and the national dish must be fufu which can be made from cassava, yam, plantain or cocoyam. It is a great shiny mass of stodge which, if I eat it, sticks somewhere just under my ribs and lies like a lump of lead for hours. The savoury soup which is eaten with it is so fantastic though that I had it twice in one day — once made with snails and goat, and the other with groundnuts and chicken.

The method of making fufu is spectacular to watch. It is easier done with two people, one pounding the mass with a very stout stick in a large mortar while the other turns it over with a clean, wet hand. In the Kumasi market outside a restaurant, near big bins full of bright pink pigs' trotters, I saw a giant fufu being pounded by two muscular men while a dexterous woman turned it, managing to avoid having her hand smashed by the two poles. Much better to watch the creation of fufu than to have to eat it! One fufu I saw prepared, and did eat, was made in Accra by Georgina Abibio who was wearing a splendid t-shirt commemorating the death of her grandfather the year before. He was aged 103.

In ingredient lists in Ghanaian recipes a regular measure is a half, or a quarter, or two, cigarette tins. A cigarette tin of light ingredients such as flour or breadcrumbs weighs a ¼ lb (250g); of heavy ingredients, such as sugar or groundnuts, it weighs ½ lb (500g). A beer bottle measures 4 cups. Nothing that can be recycled here is wasted. I bought nutmeg graters made out of baking powder tins and biscuit cutters created out of some other kind of can.

Many recipes mention garden eggs and I imagined they were some very nice eggs laid by obliging, free-range chickens under tropical vegetation and gathered in a grown-up kind of peanut hunt. They are, however, eggplants — little round white aubergines which do look just like fat little eggs.

Roast chicken with pineapple

a good sized roasting chicken

salt

4 tablespoons vegetable oil

1 tablespoon of freshly ground pepper

45g (1 $^1/_2$ oz) butter

$^1/_2$ a green pepper

grated rind of $^1/_2$ a lime or lemon

1 large fresh pineapple — peeled and cut into slices

Clean chicken and rub the inside with salt. Cut cold butter into small dice. Slice the pepper into long shreds and slice three rings of the pineapple into similar long shreds. Place the butter, lemon rind, pineapple and green pepper shreds loosely into the chicken and close the cavity with a skewer. Pour the oil into a baking pan and heat in a moderate to hot oven. Place the chicken in the pan and baste all over with the oil. Cook for an hour, basting every 15 minutes and add the remaining pineapple slices for the last 15 minutes. Watch that they don't burn. Remove from oven, let the bird sit for a few minutes before carving, and serve with the pineapple slices and vegetables.

If you prefer a *bread stuffing* here is a recipe:

1 small onion, finely chopped
1 rasher streaky bacon, finely chopped
$^1/_2$ a green pepper, finely chopped
2 slices fresh pineapple, finely chopped
1 teaspoon lemon peel, finely grated
500 g (1 lb) fresh, soft breadcrumbs
1 large egg, beaten
salt and pepper
pinch dried herbs

Place bacon and onion in frying pan and soften over a low heat. If the bacon isn't fat you may need a little margarine or oil to facilitate the sautéing. Remove from heat and add all the other ingredients. Stuff and truss the chicken as above. If there is stuffing left over, form into balls and bake in the pan with the chicken. They will take about 30 minutes to cook. Cook and serve as above.

Accra is really a collection of villages which are an amalgamation of what used to be seven branches of the Ga tribe, each of which had its own chief. One of the other major tribes of Ghana is the Fante. Today the Fante number about 1 000 000 people who live on the coast and forest fringes. They were among the first people to have encountered, and then been influenced by, the arrival of white men in sub-Saharan Africa. These people, in their unsuitable clothing, were called the *Oburinifo*, the people from the hole in the sea's horizon.

There is cultural cross-reference to Australia in one of the legacies of these foreign invaders. Among the people of Arnhemland in northern Australia the idea of flags, borne aloft by men and ships, enters both dance and visual art, probably as an influence from the Macassans who came from what is now Indonesia to search for bêche de mer or sea cucumber. The Fante had a similar encounter which has had lasting effects on their warrior groups, called Asafo.

Asafo companies were certainly in existence long before the arrival of the *Oburinifo*, but certain influences of European military organisation became obvious. The Fante were deeply impressed by the European's superior technology which, if harnessed within an alliance, could certainly be used in opposition to the mighty Asante to the north. However, they were also seduced by their uniforms, banners and flags and these are still used in creative expression even today.

There are 24 traditional states in the Fante region, which stretches along 130 kilometres (80 miles) of the Atlantic coast. Each state is ruled by a paramount chief, and beneath him are other chiefs and elders. In any one state there may be from two to fourteen Asafo companies and there can be up to seven active companies in any one town. The Fante could often fight against a common enemy, such as the Asante, but also, just as often, they could be engaged in battle among themselves. So the Asafo companies became very important in the balance of power in the state and came to be involved, not only in matters of defence, but also in philanthropic, social and cultural activities.

The most enduring of their art forms are the shrines, or posubans, these days made of painted cement. I visited one in the town of Mankessim, half-way down the road to Cape Coast. It was a two storeyed edifice adorned with gods, men and animals, painted in vivid colours. From out of nowhere came a most dreadful man, who from the pamphlet he handed me, must have been J. F. Barden, Raconteur. He insisted I sit down as he talked, or rather, shouted at great length, on the history of the nation and of the shrine and of the world at large and then he demanded an extortionate 3000 cedis in payment. He became quite menacing and I gave him what I considered a very generous 2000. Things were starting to feel rather nasty when a cutlass fell off the back of a passing utility truck and gave me, the instant crowd that had gathered, and the raconteur, a terrific fright. It broke the atmosphere, though, and I was able to continue my journey.

I had asked about the Asafo company's flags, but J. F. Barden said they were all locked up. Because of the flags' fragile nature, and the taxing climate of Ghana, there are few over 130 years old still in existence. The tradition of their creation certainly goes back to the seventeenth century and Phillips, an English trader, wrote of seeing one in 1693 with a white ground and a black man in the centre brandishing a scimitar. He described the black man as being painted, although most flags show the images as appliqued and embroidered, rather than painted.

The flags are usually three feet by five feet (1 metre x 1.6 metres) in size and made of European cotton tradecloth. Silk, satin and felt are also used. They were carried into battle and these days are borne proudly and aggressively at ceremonies of all kinds. They are carried in procession, hoisted aloft poles on the shrines and members' houses, and, most dynamically, used in dancing by specially trained dancers who are believed to have supernatural powers.

One of the flags I have seen illustrated in *Asafo! African Flags of the Fante* shows a pineapple 'tree' surrounded by a neat cane fence which

An Asafo flag from Ghana showing the obligatory Union Jack. The black oval surrounding the pineapple may refer to original Asafo shrines where a tree, inhabited by a female spirit, was encircled by a protective

suggests that it may have been an Asafo shrine as, before the construction of concrete posubans, the original Asafo shrines were trees, encircled by a fence and with a medicine mound. A female guardian spirit was believed to inhabit the tree and she is shown on several flags.

The tree for many Ghanaian groups is highly important. For the Asante, the tree and the King were sometimes synonymous and the shade of the chief's state umbrella was likened to the shade that trees, planted after a victorious battle, gave to his people. The pineapple tree on the old Asafo flag looks rather like a palm which could also refer to the former British symbol of the Gold Coast colony, that of the elephant and the coconut palm. It can also refer to the belief that for the Fante, the palm is the strongest tree and the symbol of eternity. Of course, many plants and trees on flags defy botanical identification, being highly imaginative and stylized.

Like other regalia in Ghana, the flags carry messages, warnings and proverbs. One Asafo proverb says, 'Only a brave man will go under a big tree', that is to say, venture into the unknown. Another flag shows an image which warns the unwary not to use a piece of thorny vegetation to wipe oneself after defecation. Sensible advice!

From several thousand proverbs, about 200 are used consistently in Asafo flag imagery and these can be understood and 'read' by groups all over the Fante region. Some taunt and challenge, and all identify with some specific image of power, and, in the case of the pineapple, the image of power is allied with the strength of the tree. One pineapple image shows three fruits — a large central one flanked by two smaller ones within a circle, again probably representing the cane fence of a shrine. A man is pointing to the fruit — what is he saying? I will have to go back to find out.

When we think of pineapples, what is the place with which we associate them? In a food sense, it is probably some dish concocted by a European cook and given the descriptive generic title 'Hawaiian' or 'Caribbean'. That is further geographically confused when added to such staples as pizza. In a future book to be called *'In Defence of the Pineapple'*, Hawaiian pineapple-bespattered pizzas are to be consigned to the flames of the hottest oven.

The pineapple has a modern history more closely chronicled than probably any other fruit or vegetable. This is because we have it in our history books that on 4 November 1493, Christopher Columbus was given some pineapples by some of the native inhabitants of the island of Guadeloupe. These people he called Indians because he, for the second time, mistakenly believed that he had arrived in the East Indies, in further India. Instead he had arrived in what we now know as the West Indies. My first memory of him is in one of those silly little rhymes made up at primary school that mean nothing but stick in one's head for ever:

> *"Christopher Columbus was a great man.*
> *He went to sea in a watering can.'*

I wonder if they say that in America or Trinidad or en français in Guadeloupe.

Cristoforo Colombo was born in Genoa, Italy in either 1450 or 1451, so it may be appropriate that Hawaiian pizzas do exist. Genoa was a great port and Columbus voyaged as a seaman to many parts of the known world, including West Africa. He aimed higher though and began collecting and studying maps and charts. In 1476 he and his brother moved to Lisbon where they bought and sold rare maps and then began to publish accounts of Portuguese maritime discoveries. Through his study of maps and accounts from as far back as Ptolemy in the second century, Columbus believed that the world was round and that if he was to sail west across the ocean from Europe, he would somehow reach the East Indies where boundless treasure would be his.

The Portuguese were not interested in funding his proposed voyage of discovery, so Columbus took his plans to Spain where great interest was shown. It took five long years of waiting, however, before King Ferdinand and Queen Isabella, having at last defeated the Moors, would agree to help finance the expedition which set off in three ships in August 1492.

The voyage took four months before land was sighted for the first time since leaving the Canary Islands. This was the island of San Salvador and Arawak guides showed Columbus their trade routes to Cuba and Hispaniola. Columbus returned to Spain thinking he had discovered at least some part of Asia, but not carrying a cargo of rich silks, spices, jewels and gold as he had hoped. Instead, he carried an alligator, several parrots and Arawaks, a small quantity of gold, a canoe, a few hammocks and cotton samples, and a bundle of tobacco leaves which he did not know how to use. Because his flag ship, the *Santa Maria*, had run aground and sunk off the north shore of Hispaniola, 39 unwitting seamen, who couldn't be accommodated on the other two ships, became the first European settlers in the West Indies.

Columbus made four voyages to the Caribbean but our interest is in the second because of his fateful meeting with the pineapple on the island of Santa Maria de Guadeloupe. The second voyage was jointly organised by Juan de Fonseca, a priest at Isabella's court. This time there were seventeen ships and 1200 men. There were builders and carpenters, merchants and clerks and, importantly, a number of priests to start the work of converting the heathen 'Indians' to Christianity.

The pineapple, *ananas sativus*, originated in the tropical regions of Brazil and was indigenous to the dry upland areas of the Matto Grosso. At least three wild species are still to be found there. It has been cultivated since ancient times and had spread widely over Brazil, north-east South America and the West Indies long before the arrival of the Spanish and Portuguese. The name, *anana* comes from the Guerani language of Brazil and corruptions of this word are the most common name for the fruit throughout the world today. It is strange to think that French, Hindi, Malay and many African languages all use the original Brazilian word. The Spanish, observing the resemblance to a pine cone, called it piña and the English followed suit.

One imagines that Christopher Columbus must have sampled the extraordinary fruits with which he was presented that day in Guadeloupe. Presumably they must have been at the peak of freshness and the Amerindian donors would have peeled them and offered them to those strange white men who didn't know the treat in store. I am sure Christopher did not have in mind the combination with the pizza.

What would we have done without the Spanish spreading their horticultural talents all around the new world and bringing back such treasures in the form of the potato, the tomato, corn — we can delete tobacco — and the king of all fruits, the pineapple? Corn and the pineapple have great qualities for storage so return voyages from their point of discovery to the Old World were not a particular problem. Tomatoes must have presented a much greater degree of difficulty. Queen Isabella could have said, 'O.K., I can see this plant that you call a tomato, but who says they're so fantastic. I want to taste it.' Much easier for the promoter of pineapples who could not only proffer an extraordinary and beautiful specimen but then dress it for the table for immediate consumption.

The Spaniards suffer from quite a degree of bad press owing to their rather drastic methods of subjugation of the masses and conversion to the true faith. However, the botanists and gardeners who went along with the priests, or sometimes combined botanical and spiritual vocations, were knowledgeable, gifted observers, and entirely dedicated. The Spanish colonies were flourishing gardens when many of the English settlements were barren deserts. As well as embracing the new botanical specimens they discovered, and immediately putting them to use in a wonderful variety of food, they introduced their own indispensable fruits, grains and vegetables to enhance their new colonies. Peaches, grapes, olives, onions, lettuce, wheat and barley grew happily alongside the indigenous crops. One reason for this is obviously the difference between the national cuisines of Spain and England. I know which I would prefer!

Quince & Pineapple
Ancient & Modern
Tomato Jam
Pineapple
Marmalade

Left: Jams Ancient and Modern — muffins and marmalade. The silver fruits come from Vietnam as do the jam-pot 'hats'. The print is from New York and the key-hook from Alexandria, on the Potomac River near Washington, DC

Pineapple & tomato jam–New World jam

1 large pineapple, peeled and chopped

1 kg (2 lb) tomatoes, peeled and chopped

375 grams (12 oz) of sugar to each 500 grams (1 lb) fruit mixture

Place pineapple and tomatoes in a large saucepan. Bring slowly to the boil and simmer until pineapple is soft. Measure the mixture and return to the saucepan. Add the sugar according to weight. Stir until sugar dissolves then bring to the boil. Skim off scum and boil for approximately an hour or until jam is thick. Pour into hot sterile jars and seal.

If that was New World Jam, then the next recipe should be an entry into Jams Ancient and Modern, made as it is with pineapple from the New World and quinces which are so associated with the ancients. A member of the apple family, the botanical name for the quince is *Cydonia*, from the ancient city of Kydon on Crete, where an especially good variety grew. It is also native to Persia. The ancient Greeks ate them with beef or pork and used an infusion of quinces for hair dye — it must have looked a little like the henna used in Muslim beards. It was also used as an antidote for hangovers, poisons, stomach-upsets and fevers. What a useful thing — and it was further believed that dreaming of quinces meant good fortune to come.

The one thing wrong with quince is that they are so hard that cutting up a quantity to use in preserves spells ruination for one's hands. Moreover, when it is time to deal with the quinces in my garden it is also the season for Jerusalem artichokes which require a deal of perseverance as well. However, the end results justify the effort, especially with the quinces. Their colour above all is sublime — a rich, pale ruby.

This recipe is an early one from the history of the colony of New South Wales, so the quince must have arrived soon after Governor Phillip in 1788. Because pineapples are also used, they must have arrived relatively early too.

Quince honey-jams ancient & modern

1.5 kg (3 lb) quinces, peeled and cored

2 smallish pineapples, peeled and cored

2 $^{1}/_{2}$ kg (5 lb) sugar

2 $^{1}/_{2}$ cups water

Mince the quinces and pineapples. Make a syrup of the sugar and water by boiling together briskly for 15 minutes. Add the fruit and boil quickly for 30 minutes or until it sets when tested. Pour into warm, sterile jars and seal.

Explorers who arrived in the New World after Columbus, wrote of the incomparable flavour of the pineapple, while in 1513 King Ferdinand of Spain received a letter from Oviedo with the first drawing of the fruit. Soon after this it arrived in Europe and took it by storm.

In 1566 Spain imposed a ban on foreign botanists and plant-hunters and forbade publications on the natural resources of South America. The ban didn't apply to importing the plants to Spain itself, and so the cultivation spread, not only of food crops, but flowers such as the dahlia, nasturtium and fuchsia. By 1548 pineapples had been planted and were thriving in Madagascar and by 1590 in India.

Strangely enough, the most fascinating chronicle of their cultivation took place in Europe where the climate was not nearly so kind. In today's age of air-freight and refrigeration it is easy to forget how grim and fruitless winter must have been in the north. When we lived in University-student's - family - self - imposed-penury in Norfolk for a year, we discovered it in a slight way and made do with windfall apples kept in a box in the freezing cellar.

Despite the advantages of twentieth century transport and technology, East Dereham's markets were still rather limited and cabbages and Brussels sprouts were much more plentiful and cheap than anything else. My children utterly refused Brussels sprouts until I made them into a wonderful curry. It was modelled on a recipe for curried winter greens. This must have been created out of desperation by a home-sick Indian expatriate. I also made nettle soup and created other meals from weeds in that impecunious year in East Anglia, where I learned what suburban *ennui* was all about and lost the golden illusions of Britain and the British engendered by my teacher in fifth class.

I certainly didn't buy pineapples that year but, as we had arrived in England after two years in Papua New Guinea, where the pineapples grew to a simply colossal size, I suppose we didn't miss them. I might have bought tinned except that we had a fierce anti-apartheid determination not to buy anything South African and that was the source of all the local supermarkets' supplies. In Papua New Guinea, of course, the tinned variety came from Australia and it was quite shocking to find that in the best restaurant in Port Moresby, with all the marvellous tropical fruit available at the doorstep, tinned pineapple with ice-cream featured as an exciting dessert.

Cultivation of the pineapple did not commence in Australia until the late 1840s. Of course it is most likely that some curious home gardeners had planted crowns and grown them for the table before that. The fruit that arrived in Australia initially would most likely have come from India. In 1848 Mr D. F. Gerler raised plants from crowns which had been imported from the West Indies. His plantation was at Nundah on the outskirts of Brisbane. This district was then known as the German Station as it had been the home of the early German Lutheran missionaries. The first pineapples had been grown in Brisbane but the great success was at Nundah which then spread in all directions to the agricultural areas surrounding Brisbane

These days Queensland grows practically all Australian pineapples and each year over 130 000 tonnes are produced. About 85 per cent is used for processing, principally by the Golden Circle Cannery in Brisbane whose supply comes mainly from southern Queensland.

The Big Pineapple at Nambour is a famous tourist attraction and there is a pub at Kangaroo Point called the Pineapple Hotel. I hadn't noticed it in the daylight as I passed by in buses going to and from a conference, but in the evening a splendid neon sign announced its presence to the world. It was established in 1864 which makes it one of Queensland's oldest pubs. It was named for the pineapple plantations which dotted Kangaroo Point at the time, and, as there were no bridges over the Brisbane River, it was an important staging point.

No-one can be quite sure just when the first pineapple was grown in Europe. It was probably cultivated in Holland but the first in England was grown by John Rose, Charles II's gardener. A painting recording this horticultural feat is to be found in Ham House, near Richmond. It must have been rather a miserable specimen, because more famous in the annals of the *ananas* is the fruit produced in 1714 by Henry Tellende, gardener to Sir Matthew Decker at Richmond, which was said to be of perfection.

Oranges and other citrus fruits were grown indoors — in orangeries — in the seventeenth century, but it was in the eighteenth century that more and more exotic varieties of fruit were tried, with varying degrees of success. Of course this was a hobby and a consuming passion for the rich — the poor were still eating their windfall apples stored in the cellar and the vast quantities of dried fruits, harvested in the summer, in special puddings and cakes. The home-grown pineapple became a status symbol and in the glass-houses of the rich it became a more and more common sight to see rows and rows of pineapple plants grown in individual pots. Peaches and grapes were other favourite hothouse fruits of the time but they were much more touchy and inclined to curl up and die if for some reason the temperature within the glasshouse was to suddenly drop.

In France, in the hot-houses of the Duc de Bouillon at Evreux in Normandy, there were 6000 plants, but William Speechley, who was gardener to the Duke of Portland at Welbeck, grew 10 000 plants and in 1779 published a book on the cultivation of the pineapple. *Chambers' Encyclopaedia* of 1904 suggests that it was Mr Bentinck, ancestor of the Dukes of Portland, who first cultivated the pineapple in England. The broad acres of a pineapple plantation in the tropics is a beautiful sight but it must have been quite marvellous to have seen thousand upon thousand under glass, especially when they were in fruit.

One reason the pineapple did so well is that, despite being tropical, it is remarkably tough, and it is highly resistant to disease. Pineapples are very easy to grow under the right climatic conditions. I have just stuck the tops in very indifferent soil and off they go, doing their best to please the eye as well as the tongue. I planted about 50 in my rented garden in Nairobi. I feel sure the landlady had them dragged out as soon as we left.

When pineapples had established themselves as the 'Kings of Fruit' in Europe, it must have been tempting for cooks to try and prolong the seasonal enjoyment by preserving. On the next page is an old recipe from Buckinghamshire, which would have been used in the nineteenth century.

The perfect pineapple grown for canning by Golden Circle in Queensland

Preserved pineapple

To every 500 gram (1 lb) of fruit, weighed after being pared, allow 500 gram (1 lb) of sugar and 125 ml ($^1/_4$ pint) of water.

'The pines for making this preserve should be perfectly sound, but ripe. Cut them into rather thick slices, as the fruit shrinks very much in the boiling. Pare off the rind carefully, that none of the pine be wasted, and in doing so, notch it in and out as the edge cannot be smoothly cut without great waste. Dissolve a portion of the sugar in a preserving-pan with the water. When this is melted, gradually add the remainder of the sugar and boil it until it forms a clear syrup, skimming well. As soon as this is the case, put in the pieces of pine and boil well for at least half an hour, or until it looks nearly transparent. Put it into pots, cover down when cold and store away in a dry place.'

Pineapple juice

I am not quite sure exactly what is meant by the notching in and out, but the best way to peel the fruit, after 'unscrewing' the crown, is to try and remove a thin rind of skin with a sharp knife and then go round and round in a circular spiral removing the 'eyes'. I am just as frugal as that nineteenth century housekeeper and peel the fruit over a saucepan, catching all the juice and pieces of flesh that are attached to the eyes. All the pieces of skin go in too.

Cover this would-be rubbish with water and bring to the boil. When it has boiled for a minute of two put in a few tablespoons of sugar and boil for a good ten or fifteen minutes, both to reduce the liquid a little and to bring out the flavour. Cool and strain through muslin. It is a delicious drink by itself or can be added to any other fruit juice.

Another recipe made out of what would often be thrown out is this old-fashioned but delicious dessert, Pineapple Flummery.

Pineapple flummery

1 heaped tablespoon plain flour

1 heaped tablespoon gelatine

1 $^1/_2$ cups sugar

2 cups juice made from boiling up the skins of pineapple

juice of 2 oranges

juice of 1 lemon

Mix dry ingredients well together, then add all the juices, stirring carefully. Bring to the boil then stand aside to cool thoroughly. Beat for 20 minutes until thick and creamy white and then pour into a glass bowl. Place in the fridge to set.

Preserving is a funny word. When is it a compliment rather than an offence to describe someone as being well-preserved? When he or she is 60 or 70 or 80? An Egyptian mummy is well-preserved, but I sure don't want to look like that, thank you very much.

While I am writing this section of my book, I am preserving my tree full of quinces in a wonderful variety of styles. I have just made quince jelly with a cup full of pinot noir rosé — the perfect coloured additive — and sitting on the stove, drying out, is a tray of quince paste. So many different ways of making a wonderful season's bounty last for the rest of the year.

Here is another recipe for a form of preserved pineapple. Preserved it is, but it is so irresistible that it won't be preserved for long. Hide it behind other things in the fridge!

The recipe for this alcoholic preservation of the pineapple is fun to do because one must boil up the whole fruit, minus the leaves of course. It is such an odd thing to do — it is like making *gulab jamun* for the first time, or enjoying the virtue of having saved the paper bag that one must use for sweating the skins off some grilled capsicums.

Brandy or rum pineapple

1 medium pineapple
1 clove
1 teaspoon brown sugar
450 grams (14 oz) sugar
brandy — or rum if you are living in the West Indies

Unscrew the crown of the pineapple and scrub the fruit thoroughly. A medium-sized pineapple is best as the slices will fit nicely into a wide-necked jar. Place in a large saucepan with the one lonely clove and the brown sugar and enough water to cover. Boil until just tender — i.e. able to be pierced with a skewer. Remove the fruit and cool. Retain four tablespoons of the cooking liquid. When the pineapple is cool, peel, core and cut into slices about 2 cm (1 inch) thick. Place the slices in a bowl, sprinkling the sugar between each layer and on the top. Leave on the bench overnight. Then drain off the syrup and mix with the retained cooking liquid. Place in a saucepan and bring to the boil. Boil until clear. Skim and pour into a measuring jug. Measure out an equal quantity of brandy or rum and set aside. Place the pineapple slices into a hot sterile jar. Bring the syrup back to the boil and pour over the pineapple. Add the brandy or rum, making sure the fruit is covered. Seal and leave for at least a month (if you can!).

I first made this in a hot summer and found that the cupboard wasn't a cool enough place for storage and that the explosive alchoholic mixture was threatening to errupt. I bravely defused a dangerous situation by drinking some of the liquid and placing the jar in the fridge.

In the past, and still today, the pineapple has lent its shape to all kinds of architectural detail. It is represented in roof finials and gateposts and garden decorations. In Alexandria, one of the historic ports on the Potomac River near Washington, D.C., the earliest pineapple gate-posts were formed from real fruit. When sea captains returned home from voyages to such places as the West Indies, they would bring with them quantities of pineapples. Two would be impaled upon their gate-posts as an indication of a safe arrival and an invitation for friends to drop by. Perhaps this is the origin of the pineapple as a symbol of hospitality in this part of the world. The souvenir shops in these colonial towns have quite a selection of objects in the shape of the pineapple — trivets, key-racks, brooches and ear-rings.

Surely the most spectacular architectural pineapple is to be found in Stirlingshire in Scotland. John Murray, the fourth Earl of Dunmore, was keen to keep up with competition in the cultivation of exotic fruit and in 1761 had a new hothouse constructed. Not for him a Palladian facade or neo-classisism of an ordinary kind, but rather a huge crowning pineapple for the central chamber. It can be described as a folly except that follies are usually without purpose or point and this architectural *tour de force* was certainly functional. More so these days too as one can rent it through the Landmark Trust.

Perhaps for breakfast during a stay within this Scottish pineapple one could enjoy pineapple muffins and any one of the delicious pine-apple jams.

The Dunmore Pineapple in Stirlingshire, Scotland built by John Murray, 4th Earl of Dunmore, as a hothouse in 1761

Pineapple muffins

125 grams (4 oz) butter or margarine

125 grams (4 oz) sugar

1 egg

1 cup pineapple pulp

1 3/4 cups self-raising flour

1/2 teaspoon nutmeg

1/4 teaspoon baking powder

Preheat oven to 200°C (390°F) and grease muffin tins. Cream butter and sugar, add egg and pineapple. Combine with dry ingredients and place spoonfuls into tins. Bake for about 20 minutes. Makes a dozen.

Pineapple & macadamia nut jam

1 medium pineapple — 500 grams (1 lb) of prepared fruit

75 ml (4 tablespoons) water

500 grams (1 lb) sugar

15 grams (1/2 oz) macadamia nuts — chopped

Peel and core fruit and chop finely. Place in pan with the water and bring to boil. Simmer gently until the fruit is tender and becoming transparent. Add sugar and stir until dissolved. Continue simmering until a light skin forms on the surface. Add the macadamia nuts and continue cooking until jam sets on a cold saucer. Pour into hot sterile jars and seal. Makes three cups

Pineapple & walnut jam

2 medium pineapples

6 cups of prepared fruit

6 cups sugar

juice of 3 to 6 limes, depending on size

juice of 3 oranges or mandarins

1 teaspoon grated orange rind

200 grams (7 oz) chopped walnuts

Place all ingredients, except the walnuts, in a large pan. Bring gently to the boil, stirring to mix and dissolve the sugar. Cook gently, stirring occasionally until the fruit is transparent. Just before setting point is reached, add the walnuts. When the jam sets pour into hot sterile jars and seal. Makes about six cups of jam.

This is another delicious nut pineapple combination. I always find it a dilemma just what to do first when rind and juice are required from the one lemon or orange, especially when one is faced with those smooth billiard ball fruit that refuse to cooperate with a hand grater. So when I have time I pare the rind thinly from half a dozen fruit and grate it in the blender to a lovely fineness. Then I put the lot in a plastic bag and keep it in the freezer. You can mark the bag with how many fruits were used but many recipes, such as this one, use it by the spoonful.

Pineapple honey

1 medium pineapple — 500 grams (1 lb) of prepared fruit

500 grams (1 lb) sugar

1/4 cup of lemon juice

75 mls (4 tablespoons) water

Peel and core the pineapple and chop finely. Place in pan with sugar, lemon juice and water and stir over low heat until the sugar is dissolved. Leave the mixture to stand for eight hours and then cook it over a low heat, simmering gently until the mixture is transparent and thick. Test for setting and pour into hot sterile jars. Makes about three cups of preserve.

Pineapple marmalade

4 cups of pineapple — finely chopped

3 lemons or limes, finely chopped

4 cups water

4 $^1/_2$ cups sugar

This is I think the best of all the different sweet pineapple preserves. As well as being quite sinfully delicious on the breakfast toast it can be put through the blender with fresh fruit and lemon juice for a good sauce to go with ice-cream and adds an excellent flavour to home-made ice-cream — put a spoonful in the blender with the fruit before adding to cream and freezing.

Place prepared fruit in a large saucepan. Pour on the water and allow to stand overnight.

The next day simmer gently for about 20 minutes or until the lemon rind is soft. Add the sugar and stir until dissolved. Boil rapidly for about 40 minutes or until jam sets when tested on a cold saucer. Pour into hot, sterile jars and seal.

In some of my kitchens the cold saucer has had to be kept in the fridge. Pakistan was the hottest in which I have cooked, and the kitchen was like something out of hell, both in terms of heat, function and appearance. We arrived in Peshawar in the North West Frontier Province in June and for the first three months the heat was almost unbearable and the kitchen added insult to injury. However, that wasn't the most important thing in life, and Peshawar was wonderful. The first daily excitement during those first few months was to collect the Frontier Mail pushed under the gate each morning and find what the temperature had been the day before. Had it actually hit 50°C (122°F)? Boy it was hot!

In Nairobi, where I did most of my retesting of recipes, the main problem with preserves was the supply of jars. They were few and far between in my household, and many that I had acquired from friends were plastic and not glass. Glass must be expensive there. The plastic ones were all very well, but sterilising them was a little difficult. I couldn't place them in the oven, even ones standing too close to the jam boiling away on the top of the stove melted into most unjarlike shapes. When pouring a very hot mixture into them they would get so soft that picking them up became a very dangerous procedure. A fountain of hot jam would shoot out as the jar had become something of the texture of a rubber glove! Something definitely happened to the neck of the bottle and lids would not screw on at all. So it was out with the cling wrap and give those ones away to friends.

Pineapple, pear & lime jam

1 kg (2 lb) pears — not too ripe

1 medium pineapple

5 or 6 limes depending on size (small lemons can be used)

2 kg (4 lb) sugar

$^{3}/_{4}$ cup of water

good splash of rum

Peel, core and finely chop the pears and pineapple. Squeeze the juice from the lemons and finely grate the rind. Place all ingredients except the rum in a large saucepan and over a low heat stir until the sugar is dissolved. Bring to the boil and boil rapidly until the jam sets when tested on a cold saucer. Remove from heat, add the rum and stir well. Pour into hot sterilized jars and seal. Makes about six good cups. It is a lovely dark gold colour.

Pineapple & papaya preserve

2 cups of finely chopped pineapple

4 cups of diced papaya — not too ripe

sugar

$^1/_2$ a cup of lime or lemon juice

good splash of rum

Combine the fruit in a saucepan with a quarter of a cup of water. Simmer until the pineapple is tender and the papaya more or less a pulp. Measure the cooked fruit and return to saucepan. For every cup of fruit add three quarters of a cup of sugar. Add the lime juice and bring the mixture to the boil, stirring to dissolve the sugar. Boil rapidly until thick but this preserve need not be tested for setting. It is inclined to stick, especially in the final stages of cooking, so watch carefully and stir occasionally. Remove from heat and add that good slurp of rum. Any alcohol could really be used but rum seems to go very well with pineapple. Pour into hot sterilized jars and seal. Makes about four cups.

Pineapple & apricot jam

1 medium pineapple — peeled, cored and chopped finely

250 grams (8 oz) dried apricots, chopped

5 cups boiling water

1 kg (2 lb) sugar

2 tablespoons lemon juice

Place the apricots in a saucepan and pour on the boiling water. Leave overnight. Add the pineapple and boil until the fruit is soft. Add the sugar and lemon juice and stir to dissolve the sugar. Bring to the boil and boil rapidly until setting point is reached. Pour into hot sterilized jars and seal. Makes about four cups.

The cultivation of pineapples in England remained widespread through the latter half of the eighteenth century but tailed off after about 1830 when advances in the technology of ship-building meant that voyages to and from the West Indies could be completed quickly enough to bring fruit to market in England in good condition. So the massive nurseries with their succession houses, which were built to bring on fruit at different times, became dinosaurs, and few survived.

One that did is now an impressive garden museum at Heligan in Cornwall. Its aim is to become a living museum of nineteenth century horticulture, but it dips back into the eighteenth century to look at the cultivation of the 'pine'.

The fruit were usually grown in low frames or pits and heat was supplied by decomposing manure or tan bark, sometimes supplemented from flues heated by stoves. At the Dunmore pineapple, the chimneys of the flued walls providing heat were disguised as Grecian urns. The Heligan pine pit has a series of south-facing growing chambers with glass lights. The earliest glasshouses were simply glassed-in southern walls with fruit trees espaliered on them. In the Heligan chambers the plants are grown in 20cm (8 inch) pots immersed in 35cm (14 inches) of fermenting tan bark and the front and rear walls are honeycombed with holes connecting to a rear channel filled with manure. The air that enters through the holes is warmed by contact with the decomposing manure and too much heat can build up which is then counteracted by opening the heavy glass lights. The ideal temperatures are 13°C-15°C (55°F-59°F) at night and 25°C-30°C (77°F-86°F) by day.

The early cultivars numbered about 50, with wonderful names such as Black Jamaica, Brown Sugarloaf and St Vincent. Some were huge — Trinidad could weigh up to 12 kg (26 lb) and some were prized for sweetness, such as Moscow Queen. This is one variety grown at Heligan today along with Smoothleaf Cayenne. In Australia today the Smoothleaf Cayenne and the Roughleaf are the two commercial varieties. Smoothleaf Cayenne is the main commercial variety in Kenya too because of its convenient size and shape — it is ideal to put into a can. Older Queenslanders, however, will tell you that the only one worth eating fresh is the Roughleaf.

In the United States of America, most pineapples now come from Hawaii. The history of their introduction there is unknown but it is generally believed that they were brought to the islands by a Spaniard from South America. They were cultivated and shipped to San Francisco before 1880 but the industry really took off when the Smoothleaf Cayenne variety, so suitable for canning, was introduced in 1886. Commercial canning began in 1892 and is now the islands' second largest industry, using a large proportion of the crop.

Queen Victoria in a splendid hat. A Kenyan terracotta version of The Queen Empress wears Bromeliad Melunie as her crowning glory. The real Queen Victoria was given a piña cloth petticoat in 1850 and Prince Albert, far from being forgotten, was at one time given a handkerchief, and, at another, a pair of finest piña drawers

Spanish Moss and other bromeliads

A pineapple in bloom is a pretty thing. From each of the little round segments protrudes a small purple flower and one can see the similarity to those pretty bromeliad potplants one can buy, with a fan-like flower, half like a feather, half like a leaf, in bright pink or red or purple.

The pineapple is the best known member of the large family Bromeliaceae, a group of highly ornamental plants largely from the rain and cloud forests of central South America. Another cousin is the strange dead-looking Spanish Moss which seems about as far removed from the pineapple as it might be from a melon. These variations in resemblance occur because of the wide range of ecosystems in which they grow, and to which they have adapted. Their distribution is from Virginia and Texas at 37°N in southern United States of America to the central parts of Argentina and Chile at 44°S. They grow in elevations ranging from sea level to over 4000 feet above sea level.

Some species can withstand near-freezing temperatures, while others can survive the blistering heat of desert regions. Others again have root systems which are reduced and serve the primary function of anchorage though in some species the functions of water and nutrient absorption are retained. Species found in arid areas appear silvery-grey owing to the dense concentration of little modified 'hairs' or trichomes on the leaves which help to reflect sunlight, thus reducing heat and water loss. Some of these varieties look like strands of wire or stiff grey string tied in knots. Some are tiny and may be less than 3 cm (1 in) tall while others are huge, like the columnar giant of the Peruvian Andes which may grow to more than 9 metres (9 yards) in height.

Bromeliads, in particular the epiphytic species, conserve water by having their leaves arranged tightly in a rosette to form a water-holding tank or vase. They look to me like little rock pools with a 'sea anenome' poking up above the surface of the water. Some atmospheric species grow happily on telephone wires. We frequently saw these 'air-plants' in Trinidad.

The origin of the name Bromeliad is attributed to the French explorer-priest, Charles Plumier. He chose it in honour of the Swedish botanist Olaf Bromel, for a plant he encountered while exploring in the Caribbean. It is a large family of 2500 species and 51 genera in three sub-families.

In the Singapore Botanic Gardens one can visit a splendid bromeliad collection. In a special enclosure there are over 300 species and more than 500 hybrids — 19 000 specimens in all which were acquired from the Shelldance Nursery in the United States of America in 1994 and given to Singapore by Lady Yuen-Peng McNeice. It was a marvellous surprise to me as I had no idea it was there, within the National Orchid Garden, which I had come to visit.

There were so many beautiful varieties — bright fire-cracker red flower spikes rising from greyish-green leaves, variegated leaves of cream, pink, green and grey, huge silvery leaves with sharp serrated thorny edges, pink tips, scarlet hearts, lipstick-bright leaves, some like flat pink lotus, others suspended from the roof like faded Christmas decorations hanging in an abandoned country store. It was wonderful.

There are no recipes in this book using canned pineapple. How could a tin of golden rings compare aesthetically with a fresh one? And many of the recipes one finds using the tinned variety are to be put into the same category as the Hawaiian Pizza. Caribbean is another adjective to watch in the quest of what is fair or unfair to pineapples. However, at least it is a reminder of Cristoforo and Guadeloupe and all that. I have to admit too that if you come across a quantity of crushed pineapple in a recipe, it just might be easier to open a can of that sort, though, of course, I wouldn't. There was though, something very aesthetically pleasing in the vision of tins that I had in Kenya.

Early one morning I drove out to Thika from Nairobi, a 45 minute journey. If anyone outside Kenya has ever heard of Thika it is usually because of Elspeth Huxley's charming story of her childhood, *The Flametrees of Thika*. Her parents had gone there as pioneers and Elspeth, obviously a very observant little girl, grew up in a rather haphazard sort of way.

Agriculture in Kenya was fraught with a multitude of problems, with pests and diseases that the British settlers had never encountered before. Now of course, most of these problems have been overcome and Kenya produces an enormous number of primary products. The most appealing are the flowers, and while I lived there I would buy every five days or so huge armfuls of roses, lilies, alstroemeria, lisianthus, Queen Anne's Lace, daisies, statice, gladioli, gerberas and practically anything you could think of. Most were grown under plastic near the shores of Lake Naivasha, so the beauty was not in the field. It was such a difference from many of the places in which I've lived where flowers were absolutely non-available, or if available, then non-affordable.

One 'flower' that wasn't available, was the tiny pineapple on a long stalk which exclusive florists all over the world now stock. This wild variety is Ananas anasoides and is decorative only. It does have some useful resistance to soil nematodes. The roots have hairs and are more difficult for the nematodes to climb. This variety is therefore used in cross-breeding.

In an old lithograph from the *Bengal Civilian* in 1853 there is a rather curious depiction of 'A Woman of Cheribon'. It is a stylized backview of a Javanese woman who looks rather more African than south-east Asian. Tucked into her shoulder cloth is a basket of pineapples. They are tiny little ones the size of large plums.

To Thika then I went, in order to visit the plantation and factory of Del Monte Kenya Limited. Mr Brian Megilley, the Director of Agricultural Research and Training, was a very charming and interesting man who had been involved for 32 years in growing pineapples, 26 of them in Kenya. His interest had started at college and he spent a year at the St Augustine campus in Trinidad. We had a cup of coffee and a chocolate chip cookie and he told me of a salad recipe using the leaves of the pineapple. In Thailand young pineapple leaves are included in curry in the same way as bamboo shoots.

A woman of Cheribon with a basket of pineapples — from the Bengal Civilian, 1853

Pineapple leaves vinaigrette

Take the white parts of the crown leaves and par-boil, drain and marinate in a sesame oil vinaigrette for several hours. It is a little like globe artichoke. Your common sense will tell you which tough bits to discard. A useful piece of advice is to wash very well, not only for dirt but, in some parts of the world, for pesticides.

In Trinidad, almost the entire crop of pineapples growing in our perpendicular back garden was stolen. In the Caribbean, theft of food crops is a particular problem and is termed praedial larceny. Apart from loss of income to the farmer, there is a health problem that was revealed when we were living in Port-of-Spain. Cabbages had been stolen and then taken to market but what the thieves did not know, or care about, was that the crop had been sprayed with a very strong pesticide which needed several days to become neutralised. Fortunately nobody died from eating the contaminated vegetables, but several people became ill. In Kenya this praedial larceny is known colloquially as 'African finger-blight'.

Thika is a perfect place for the cultivation of pineapples for canning, in terms of both altitude and temperature. Coastal pineapples are not so good for canning as they have a large core, caused by high temperatures, which means the fruit is sweeter. This, of course, is wonderful for eating as fresh fruit, but canned fruit shouldn't be too sweet as when the acid reduces, botulism can result. Fruit which ripens in the cool season is higher in acid and the best season in Kenya is October to November, and again January to March. The altitude at Thika is 4900 feet (1600 metres) above sea level.

It was a beautiful day, with sunshine and blue skies, but rather spoiled by the most terrible smell hanging over the factory and offices. I had expected a sugary, pineappley smell, rather like the marvellous aroma put forth into the universe by the rum factory near Georgetown, Guyana. This was unfortunately much more reminiscent of muck-spreading in Norfolk. I recalled some childhood book where Nanny admonished her charges to 'Spit, children', whenever within sniffing distance of a nasty farmyard smell. This was far worse than any ordinary pastoral pong and came from the waste water from the cannery. It goes into four lagoons in the treatment plant before returning to the river and the cool nights at that time of the year do something to the bacteria which makes it so odoriferous.

We set off to see the fields in a four-wheel drive vehicle, and once away from the cannery the smell vanished. 30 000 plants grow to an acre (74 000 to the hectare), and this plantation was 13 500 acres (5500 hectares) in size. The botanical name of this pineapple is *Ananas comusus*. The fruit is grown from crowns, suckers and slips. The first crop at Thika is from crowns, which grow to a much more uniform height than the second, ratoon crop, which comes from suckers.

Harvesting a field of pineapples at Thika in Kenya

Brian said that the pineapple is the champion of women's 'lib' as it does not need a male. It produces fruit without fertilization, and therefore has no seed. Seeds can be produced by pollinating with a paintbrush, but then the fruit is rather unsightly with black seeds or marks and would be unattractive in a canned form. It is a multiple or compound fruit, meaning that each little segment has ovules and a pistil and is, in fact, an individual fruit.

The pineapple crowns are planted in holes made in long strips of black plastic. The plastic warms the soil and retains the fumigant which is applied. Pulp from the factory waste, of which there is 48 000 tonnes a year, is used as organic fertiliser. It has by then been squeezed dry of every last drop of juice which is used at most times of the year as the canning liquid.

When the plants are eleven or twelve months old, a slurry of china clay and water with ethylene gas is sprayed over the crop, at night, to change the plants from a vegetative to a flowering state. Forty acres (16 hectares) are sprayed in a night and the application of the slurry to different areas means that the crop can be staggered to ripen at different times. Pineapples respond well to growth regulators — they are easily manipulated — another endearing quality! Perhaps they are in need of a certain protection, as I have recently read of drugs being smuggled into Australia in a tin of Chinese tinned pineapple, and not so long ago a certain woman in northern New South Wales drugged her fellow workers with a pineapple cake laced with cannabis.

The growth cycle of the pineapple varies from as short as fourteen months in Cameroon or up to four years in parts of the United States of America. In both Kenya and Australia the first fruit takes an average of about eighteen months and the second, ratoon crop, another twelve or eighteen months.

At Thika, one sucker per plant is left on the plants after the first harvest for the second crop and six months after that, another night spray is performed. Evaporation is too fast during the day. If growing a pineapple in a pot a rotten banana thrust into the heart will bring on a flowering state—obviously a state of shock at the immaculate conception. I tried this technique to coax a reluctant orchid and only attracted a trail of ants.

At Thika, 280 000 tonnes of pineapples are produced per annum of which 99 per cent is exported. The harvest is carried out by hand and the workers walk behind a tractor pushing a truck. They pluck the fruit, 'unscrew' the crown and place it on one conveyor belt and the fruit on another. These are then carried along and fall into separate bins on the truck. The crowns will then later be returned to the fields to be planted while the fruit goes to the cannery.

I wasn't permitted to take photographs in the factory, but the whole canning process was fascinating. Fruit arrives at the cannery by truck from the fields and the full crate weighs 9½ tonnes. It is poured into a 'river' — a form of water transport, and then goes into an elevator, ascending the calm outside wall of the factory, before all the violence inside.

A machine called a Ginaca, named after its Hawaiian inventor, peels and cores the fruit and the discarded skin is squeezed for juice. The now hollow cylinder of fruit then goes into a slicer. The top end is the choice and the bottom is the fancy or No. 1 grade. A further sorting makes sure that fruit of the same colour makes up a whole tin. Piped hot juice goes into the cans which are then sealed. This section of the cannery is full of women standing over conveyor belts of the golden fruit and silver cans. The colour is wonderful as the women wear white coats and head scarves and bright blue gloves.

Kenya was where I started writing about the pineapple, perhaps as a defence mechanism. My husband was working at a very arduous job in Somalia with a difficult group of men. Everything seemed to go wrong while we were there. However, I made a number of wonderful friends and saw some extraordinary sights, such as the migration of the wildebeest and zebra at the Masai Mara. Hundreds of thousands of animals on the golden plain — the zebra so elegant and the wildebeest so plain, only their mothers could love them — lions and hippo and elephant and the stately, mute giraffe.

Another extraordinary species was the 'barking white matriarch' which seemed to abound, mostly in an urban environment. The genus, strangely seemed to be able to transcend certain orders, so that Nairobi barking matriarchs didn't necessarily have to be white. It seemed almost to be a contagious condition. Stay there too long and you might bark, too.

I joined the Horti, a.k.a. the Nairobi Horticultural Society, in order to borrow books from its library. Private gardens were open from time to time and we were able to visit a most marvellous one on the Old Naivasha Road with especially dear horticultural friends visiting from Canberra.

I volunteered to help with the teas at one of the Horti shows, which had two distinct benefits—to help in my pursuit of the pineapple and in furthering my field observations of barking white matriarchs at work and play. I arrived at the tea tent well in advance of my time-slot, which was 9:00am, the first for the day. Other helpers were there but I was told that the female major-domo whose name was Toots, never arrived before 10:00am. We got cracking and did all the setting up. The enamel tea-pots were huge and very chipped. The time-honoured, and it must be said, effective method of insulating them, keeping the tea hot for as long as possible, was to cover the row of pots in a thick bath towel. This towel was, or had been, beige, and may well have been clean, but it was so stained as to look not only utterly revolting, but a distinct health hazard. I suggested we just might cover it with one of the checked gingham table-cloths and that was taken as a very good idea.

Everything went smoothly until a short but formidable woman of about 70 arrived well after 10:00am. This was Toots. She nodded in the direction of the helpers and I was introduced *en passant* as she strode into the shed behind the tent. I had just been in there, cutting up a banana cake. Soon there was a furious roar, much more than a bark, 'Who has been cutting up this bahnahnah cake? The slices are far too large.' I couldn't deceive the barking white matriach and so I admitted responsibility but I added, 'Never mind, I brought a big tin of macaroons that you didn't expect.' Cranky old thing.

Many of the barking white matriarchs belonged to a women's club that published a newsletter and in its first issue for 1995, a page of pineapple recipes was included. The plucky and dependable dishes require the tinned variety of pineapple, but I have experimented and adapted one cake, the recipe for which is on the next page.

Barking white matriarch pineapple cake

125 grams (4 oz) margarine

275 grams (9 oz) sugar

2 eggs slightly beaten

500 grams (1 lb) self-raising flour

pinch of salt

1 teaspoon cinnamon

500 grams (1 lb) drained, crushed pineapple

Topping: 3/4 cup chopped walnuts

3/4 cup brown sugar

This is a rather heavy cake which is appropriate to its origins. It could be an alternative to pineapple upside-down cake and eaten as a pudding, particularly as the mixture that is meant to be the topping seems to partially descend every time I have made it.

Cream margarine and sugar. Add the eggs. Fold in sifted dry ingredients and then mix in the pineapple. Pour into a lightly greased 20 cm (8 in) springform tin. Mix together the walnuts and brown sugar and sprinkle on top of the cake. Bake at 180°C (375°F) moderate oven, for about 45 minutes. Let the cake sit on a cake cooler for 10 minutes before removing from tin. My topping fell into the cake and became a bottoming but it was not at all bad, and I suggest serving as a dessert with ice cream, yoghurt or whipped cream. It keeps very well for weeks if stored in the fridge.

Pip Robertson lent me the 'BWM Newsletter' and some other interesting books including one produced by St. Andrew's Church Woman's Guild, Nairobi and first published in 1928. It is a good old book, like the *Country Women's* or the *Commonsense* or the *Presbyterian* in Australia. But most of the pineapple recipes are fairly disgusting — stodgy nursery puddings and the like. Unfair to pineapples to mix them with tapioca, I say.

The best book of all was *Good Living, A Practical Cookery Book for Town and Country* by Sara van Buren Brugiere, published by Putnam's in 1890. Pip bought it coverless, in an antiquarian bookshop in Melbourne. There are several interesting pineapple recipes, and also written in the owner's pretty copperplate hand on a back page, is a recipe for melon and pineapple jam. She was obviously interested in cake-making as there are recipes cut out of old newspapers and others written out for some very complicated Victorian cakes. Someone in her family must have just loved cocoanut [sic] ice as there are two recipes for that, involving the boiling up of much milk. I made coconut ice by the tonne in Canberra at different times — for my own profit sometimes and for various charities at others. It is much easier these days with condensed milk.

There are strict instructions in this book for how to prepare pineapple for table.

To prepare pineapple for table

'Be sure the pine is perfectly ripe. Pare it from the top down. Remove the eyes. Hold the top in the left hand, and with a fork held in the right, scoop the fruit off from the core in pieces about two inches square. Place these in a glass dish; sift with powdered sugar and set away in the ice-box until required. Never cut a pine-apple with a knife in slices through the core; it is much less digestible, and its flavor is injured by contact with the knife. When properly prepared core and top-leaves should remain in one piece.'

With Joe Emmett in his Nairobi garden and his prize-winning pink, variegated pineapple, Ananas Bracteatus Striatus

Inside the hall at the Horti show, after my stint at the teas, I looked at all the exhibits and was delighted to discover that the winner of one section was a magnificent *Ananas bracteatus striatus* — a splendid pinkish pineapple growing in the centre of long variegated pink leaves. I didn't have my camera with me, so I took down the name of the owner/grower, Mr Joe Emmett, and wrote to him the next day. Kind man that he was, he rang me and graciously invited me to his extraordinary garden. I visited it with our horticultural visitors, Geoff and Rhondda Cleary, who had incidentally arrived with wonderful information about the Heligan pineapple project in Cornwall.

The infant Ananas Bracteatus Striatus in a pineapple pot growing happily in a Nairobi garden

Joe is a builder but also a passionate and gifted gardener specializing in orchids, cycads and cactus. His garden was full of the most curious things and the bush-houses contained hundreds of unusual orchids. The pink pineapple was looking splendid and he even gave me some babies which I left in Nairobi with some Australian friends, who are keen gardeners. They live in the suburb of Karen, named after the writer, Karen Blixen, author of *Out of Africa*. Her house, now a museum, is not very far away. The variegated babies are thriving under Sally's care.

The only other really pink example I have ever seen was in a painting owned by Major-General William Farquhar, the Commandant and Resident of Singapore from 1819 to 1823. Prior to arriving in Singapore he had been in the settlement of Malacca, on the Malay Peninsula, and during his years there had commissioned and compiled a collection of natural history drawings, of birds, fish, animals and plants. The pink pineapple was one of the 477 drawings which eventually found their way back to the National Museum of Singapore in 1995. The artist overcame the problems of revealing the fruit by erroneously placing the leaves on two sides of the stem. The upper leaves and the fruit are painted in a vivid pink so, perhaps, they were the same species as Joe's.

Pineapples appeared in many artistic forms in Kenya where traditions on the coast are very different from the interior. Foreign cultural influences from, in particular, Arabia and India, have long been felt and have assimilated with local

Major-General William Farquhar's pink pineapple from Malacca c.1817. This painting is one of 477 now in the collection of the National Museum of Singapore

African traditions to form the distinctive Swahili culture.

The history of the Swahili coast is not unlike that of Ghana or other parts of Africa and Asia, where the European powers, in their relentless pursuit of power and possessions, arrived and fought the local rulers and then fought each other. The Arabs mentioned Mombasa as early as the twelfth century as the seat of the King of the Zenj, or Black Africans. The Portuguese arrived in the early sixteenth century and attacked the town in 1505. They were determined to wrest the monopoly of maritime trade in the Indian Ocean from the Arabs, and they constructed the massive Fort Jesus in 1593. Their fortunes waxed and waned, however, and, in 1729, they were defeated for the last time. The costly, and unsuccessful, counter-offensive they launched, involved the entire military resources of the vice-royalty of Goa on the other side of the Indian Ocean.

The Sultan of Oman moved his capital from Muscat to Zanzibar in 1832, and the sultanate nominally controlled the whole of the Swahili coast, a 16 kilometre (10 mile) strip under the British Protectorate, from Kilwa, in what is now Tanzania, to the Somali border, until Kenya achieved independence on 12 December 1963. Omani additions to Fort Jesus contrast to the original construction by an Italian architect, Joao Batista Cairato, who had designed much of the important work of his time in Portuguese Goa. So we see how another of Christopher Columbus' countrymen linked up with Iberia in the discovery of the New World.

On the island of Lamu to the north, and in the great port city of Mombasa, the walls of houses and mosques built by the almost entirely Muslim population, were traditionally made of faced, coral-rag blocks. Wooden floors were supported by mangrove poles and the roof was of makuti or thatch, now increasingly replaced by corrugated iron. Carved shutters for the windows allowed both privacy and ventilation and ornate plasterwork inside the houses created shelves and niches to hold the possessions of the householders. One of the most beautiful features, and the ones that most frequently remain, despite attempts at modernising, are the intricately carved doors and lintels. These are also to be found in Zanzibar.

We spent Easter at Lamu and daily wandered through the twisting alleys on our way to Shela Beach. When I had booked our holiday the travel agent had said that we could stay in Lamu town or at the beach. She said, 'Well it's a long walk to the beach or you can catch a dhow'. Catch a dhow to the beach!!! How could one resist that? So we chose to stay in a quaint traditional house in the town. It was four storeys, built around a central courtyard. Our room was right at the top, with a view across the channel to Manda Island. Magenta bougainvillea had climbed up from the courtyard and poked its fronds through our unglazed windows. Gentle breezes wafted in muezzins calls to prayer and lizards occasionally made small scratching noises in the thatched roof overhead. Now and then a donkey protested its presence in the laneway below, but the peace was undeniable.

Peace comes in other ways to other creatures in Lamu. There is a donkey sanctuary on the pathway next to the waterfront and further inland, on a lane leading up to the pretty Swahili House Museum, there is a cat refuge as well. Donkeys and cats abound in great numbers. While listening to 'Anything Goes', one of those timeless BBC request programmes, with a presenter who sounds so invincibly British one starts to think there still is an Empire, I heard a request from an expatriate Britisher. It came as a dedication to the cats of Lamu and was Rossini's 'Duetto Buffo di Due Gatti' or 'Comic Duet for Two Cats'. It is a work for two high voices and piano and is very dear to my heart as I sang it, with a wonderful Japanese friend, in Vientiane at a fund-raising cats concert called 'A Purrfect Evening'. I have also sung it to several of my pet cats in different countries and their reactions have been startling. Dr Doolittle has nothing on Rossini.

There are no cars on Lamu. Well, there is one. It belongs to the senior Government official and we did see it, but it was stationary. So one walks, avoiding the donkeys and the cats, or one can take a dhow. We took a dhow from the town wharf to the beach one day, and another to sail over to the ancient ruins of Takwa, but as our days were spent mostly lying on the sand and flopping about in the sea, and the biggest decision was what we would eat for dinner in the evening, we felt that a little exercise was a very good thing.

The skill of carpentry and wood carving still continues in Lamu and there are a number of workshops which turn out carved beds and chairs and small objects for sale. Hotels along the coast and in Nairobi often feature the furniture in their decor. The old doorways with centuries of patination became the object of my research, but in the whole of Lamu I found not a single carved pineapple.

A carved doorway in the old Fort area of Mombasa on the Kenyan coast

Block print pineapple design muslin from Kathmandu, Nepal. Cloth such as this is used for mattresses, quilts and other home furnishings

Carved pineapples on an old chest from Kutch in India

In Mombasa, in the old Fort area, sadly many of the old buildings have either disappeared or have suffered desecration, but it is still easy enough to find beautiful doors and also balconies, support brackets and other architectural features which are as decorative as the doors. A profusion of vegetal forms twine and curl, reflecting the Muslim rules of what can and cannot be portrayed in art. Some forms I probably have to agree are bunches of grapes, rather than the pineapple, but I did find one. It is shown with leaves at both ends of the fruit, which is the way it is also carved on the woodblock which had been used to print some Nepalese muslin I bought in Kathmandu some years earlier.

Despite the fact that I was rapidly succumbing to either malaria or dengue fever, we continued the search for pineapples in Mombasa. A long walk took us past the fresh fruit and vegetable market where I was able to buy bags of smelly dried prawns. They weren't quite as dry as I like so I dried them off further in the sun, under fly covers, once back in Nairobi. I used them in a Thai recipe using pineapples which has a fascinating name, 'Galloping Horses'. I asked my son, James, who speaks fluent Thai and lives in Bangkok, if he would find out the reason. I described the dish and he said, 'Well, it probably looks as though a horse has galloped over it, Mum'. Such a wit.

Ma Ho – Galloping horses

1 large pineapple

4 cloves garlic

2 tablespoons coriander including root and stalk

1/4 teaspoon ground black pepper or green peppercorns

4 tablespoons vegetable oil

2 tablespoons fish sauce

4 tablespoons palm sugar (raw sugar will do)

3 tablespoons roasted peanuts, coarsely chopped

250 grams (8 oz) minced lean pork

150 grams (5 oz) prawn meat chopped — or a small tin, 185 grams (6 oz) of tiny prawns, well washed and drained

dash chilli sauce

2 red chillis, cut into fine rings or strips for garnish

coriander leaves for garnish

Peel the pineapple and remove the eyes by cutting spirally around the fruit. Cut into slices about 2 cm (1 inch) thick, and arrange on a platter. (If you prefer, you can briefly sauté the slices of pineapple in hot oil. If the dish is to be eaten in the winter it somehow seems more cosy!) Place garlic, coriander, pepper and vegetable oil in blender and blend until fairly smooth. Pour mixture into a frying pan and cook over a medium heat until garlic morsels turn golden. Then add pork, prawns, peanuts, fish sauce, sugar and chilli sauce. Stir fry until well cooked, dark in colour and quite dry. Cool. Place spoonfuls of this delicious mixture on each slice of pineapple and decorate with a sliver of chilli and a leaf of coriander.

Serve at room temperature with a bowl of steaming aromatic rice. This will serve about twelve people, depending on the number of slices of pineapple, how far the topping will extend and the greed of participants in the feast.

We found ourselves in Biashara Street, Mombasa, where a lot of dry-goods shops are to be found. In Nairobi these shops also line a street with the same name. What I was looking for here were kangas, the Kenyan version of a sarong. There is no particular national dress in Kenya, which is a source of disappointment to many women. For formal occasions, some will wear marvellous colourful outfits that have their origin in Senegal or Ghana or Nigeria. Regularly in the Sunday newspapers there are articles with various suggestions as to what ought to be adopted as the Kenyan national dress. In the streets of Nairobi most women wear western dress. Out in the countryside some tribal groups, such as the Masai, wear a distinctive costume of draped cloth and intricate beaded collars, necklaces, bangles and pendulous earrings. At home most women wear skirts and blouses and the ubiquitous kanga.

Kenyan women wearing pineapple-patterned kangas at a market in Nairobi

Its origin is from the Swahili coast in the mid-nineteenth century. The idea had started in Zanzibar, where ladies sewed up rectangular cotton handkerchiefs into one big piece large enough to cover the entire body, head and all, leaving only a slit for the eyes. This garment was called a leso and soon the pieced cloth became a single piece, manufactured initially in India and China, but later in Kenya and Tanzania. The earliest designs had a border pattern and white spots on a dark background. The spots resembled the markings on a guinea fowl's plumage, known as 'kanga' in Swahili, and this word has become the name of the garment.

Gardening in a pineapple design kanga in Nairobi

The pretty guinea fowl, of which there are several types, abounds in Africa. When I helped hang the Kenya Museum Society's Art Show in Nairobi in

The guinea-fowl gave the Kenyan sarong its name. Kanga is Swahili for the spots on the birds' plumage. These ceramic fowl came from Brazil, the home of the pineapple

1995, I was struck by just how many times the bird appeared in paintings. It appears in figurative form, rather than the abstract 'kanga' dots, in Ghana wax-cloths too, where, according to the size of the bird, the design was known as 'Big Turkey' or 'Small Turkey'.

Tie-dyeing of cloth is one of the most basic of textile arts and is practised in many parts of the world. In India the technique is widespread throughout the country and has been in use for centuries. The earliest pictorial evidence is on the walls of the Ajanta caves which date from the sixth century, but it is possible that the tradition could stretch back to the Indus Valley civilisation, which flourished between 2500 and 1500 BC.

The name in India for the technique of resist-dyeing, tying off areas of cloth in order not to receive dye, is *bandhani* or *bandhej*. This comes from the Sanskrit, to bind. There is an enormous variety of bandhani in India and Pakistan today — huge sunbursts of contrasting colour or myriad tiny dots forming both figurative designs and abstract patterns. Sometimes one can buy the cloth with all the threads that have tied up the undyed dots still intact. At the great craft mela in Delhi I watched fascinated as vendors from Bhuj in Kutch sold piece after piece, ripping each open so the purchaser could view the finished effect. It was a family weekend and I heard one poor fed-up husband say fiercely, 'I do not want to look at any more cloth!', as he pacified the equally fed-up offspring.

It is not hard then to find the derivation of the word bandana — the spotted or figured scarf worn by cowboys or prizefighters or sailors or workmen or young people of fashion from time to time. Of course, it comes from bandhani and was a major item in the huge export of cotton cloth from India. The word 'bandanna' originally referred to silk handkerchiefs which came from Kasimbazar, up-river from Calcutta in Bengal, and which the Dutch and English East India Companies started exporting in large quantities in the late 1720s.

Then the word became synonymous with any handkerchief or rumal, but it wasn't just a dainty square into which one blew one's nose. That was a pocket handkerchief. The one we are interested in was colourful and a sensible size of about 60 cm or 90 cm (2 or 3 feet) square; strong enough to wrap up a running-away-from-home bundle, or to tether a dog. Of course, they could be used for the nose and, as snuff-taking was a habit at this time, it was important that the colour would camouflage stains. As they became more and more common and useful, delicate hand tie-dye was replaced by choppa or printed designs, usually spots or floral sprigs.

For ease of production and packing they were produced in lengths with several squares in a rectangular piece of cloth. Each square thus had two selvedges and two sides which had to be hemmed after they had been cut. Can you see where I am leading you? Surely the original spotted kangas were made of such stuff as this —

The motto on this Kenyan kanga or sarong says in translation 'If you eat a pineapple you cannot eat any other fruit'

by the mid-nineteenth century of factory-made cloth — but still from India, and not copies from Manchester or Glasgow.

Other lasting reminders of the Indian connection are to be found in some popular and regularly sighted designs used on kangas. These are a pair of mangoes, and the Taj Mahal, both of which look really stunning on a broad backside.

In time, Swahili proverbs were added to the designs and women, if they own a few kangas, may wear one which fits their mood of the moment. I bought three in Mombasa, all featuring pineapples and still showing the dots of the original bandanas. One, in red, green and yellow says in translation, 'Although you are blaming me, to someone else I am blameless'. Another, in yellow and green says, 'I'm not proud of what I have — it is a gift of God'. Both were bought on Easter Day. The other refers to the pineapple and says, 'If you eat a pineapple you can't eat any

Kenyan kangas with a design of pineapples

other fruit', which the shop proprietor told me did not necessarily have anything to do with the fruit, and really referred to the wearer.

All kangas are bold in colour and design and are often sold in pairs which have to be cut, just like the bandanas, from which they originate. They are as long as a person with arms outstretched and wide enough to cover from neck to knee and, like the Ghanaian wrappers, perform many different functions and can be worn in dozens of different styles.

Borrowings in textile design occur not only from one country to another but within a single country. What is a single country after all? Very often it is the intervention of an invading colonial power that unifies previously autonomous states. Such was the case in India.

Several years ago I attended the All India Craft Council Annual General Meet in Hyderabad. I was extremely lucky to be there — I went with my great friend, Jasleen Dhamija, from whom I have learnt so much about textiles, and life. She is a foundation member of this Council, but because of her amazingly busy, world-wide commitments she isn't always able to attend the annual general meeting. This year she wanted to have her say on the topic for general discussion which was about children working in the craft sector. She didn't want the possibility of patronising, middle-class views of 'nimble fingers' and tradition and inheritance to endanger the rights of the child to education and a healthy life. She knew better than many in what appalling conditions some children have to work, having completed, amongst other things, a survey for the International Labour Organization on the hazardous glass industry in Ferozebad. Perhaps it is partly a result of having attended this conference that I have been a more active member of Defence of Children International and do what I can to raise both funds and awareness.

We stayed with Suraiya Hassan, an old college-friend of Jasleen's, in a wonderful oasis on the edge of the city. Suraiya, who has a most kind and beautiful face, runs a handloom factory where many old textile techniques are being revitalised and saved from extinction. She has established a primary school for the children of her weavers and others from the local village. Nearby is the tomb of a Muslim saint which is much visited by devotees. There were fields of jasmine plucked in season for the production of attar, little children who arrived quietly after dinner to have their reading heard, an amazing pack of Alsatian dogs, and some very old, rather dotty, ladies who were revered family members. One of them sat on her charpoy and did patchwork for hours on end. I felt as if I was within the pages of a South American novel, or more likely something rather surreal by Salman Rushdie.

Subhas Chandra Bose, the great Bengali patriot, was the uncle of Suraiya's late husband, and there were photographs of him in the house as well as some good modern sculpture created by a nephew. We ate wonderful food, and each morning I rose early and went for a walk with a cousin who lived in another house in the compound. The sun came up over the great rocks that surround much of Hyderabad as we pounded along the road, passed by farmers in carts drawn by pale cream oxen.

The tombs of the Qutb Shahi Kings, dating from the fifteenth and sixteenth centuries, were just a kilometre or two away, and I walked there late one afternoon, when the sun cast long golden fingers across the undulating ground. In the village women were queuing at the ration shop for rice and sugar, drawing water from the well and buying tomatoes, onions and brinjals ready to cook for the evening meal.

Before the conference began Jasleen and I explored part of the old city which she knows, like so many other cities, very well. We looked at pearls for which Hyderabad has always been famous, and bought essential oils from an attarwallah. I chose khus which is perfumed with an aromatic grass. There were great barrows of pineapples for sale in the busy streets and we had a marvellous dosai for lunch. Then we plunged into some back streets and Jasleen found, as if she had radar, the most stunning example of Chinese-style embroidery from Kutch. It was perfect.

At the end of the conference we had special dispensation to visit the Nizam's Palace, and one night we went to listen to ghazals in a beautiful garden. You can see how one can fall in love with India.

I had been to Hyderabad some fourteen years earlier with my mother and my children, Jamie and Emily, aged 10 and 9. We were rather in a state of disarray and this was the only visit I have had to India where things really did go wrong. However, we regrouped in Hyderabad in the cocoon of the nice Rock Castle Hotel where we nursed stomachs back to a state of equilibrium.

I bought from Suraiya some metres of kalamkari cloth. This is the form of resist decoration of woven cotton cloth practised in Andhra Pradesh. Designs were traditionally applied with a pen or 'kalam', hence the name. Printing is also done with woodblocks and sometimes there is a combination of the two. One can buy huge and wonderful kalamkari paintings but I would have a big problem living with the smell. There is often a prevailing and quite nauseating sourness and this is due to some dyeing processes using rice water and buffalo milk.

Kalamkari cloth printed with pineapples from India, forms a background for this painting from Haiti by N. Morin

Red and black Kalamkari cloth printed with pineapples from Hyderabad in India

My purchase from Suraiya didn't have this olfactory drawback. It is red cloth with a design of black pineapples, large and small. The table we had in Nairobi was the wrong size for any of my existing table-cloths, so I cut some of this lovely kalamkari and used it as a cloth. I later discovered in a book on block printing and dyeing of Bagru near Jaipur in Rajasthan, a very similar design. Pineapples would have grown far more happily in Andhra than in the desert state of Rajasthan. Which one came first? Did pineapples from the Portuguese colonies further south find their way to the Rajput kingdoms? Or did the English grow them in Madras? And who thought of using them as a design for cloth?

Perhaps the design itself migrated. There is a Saurashtran community in Tamil Nadu which was induced in the early seventeenth century to migrate from Gujarat in the west because of their skills in silk weaving, particularly in brocades. The splendid textiles they produced were for the local priveliged elites and were quite different from the kalamkaris. These painted cloths, made in a number of Indian locations, became more and more popular. The Portuguese called them pintadoes. They were later known as chintes from the Hindi, meaning variegated, and this would be applied to both printed and painted cloth and become the English 'chintz'. If you search, you can find pineapples on some old chintzes.

An African American friend in Kenya was researching her doctoral thesis on clothing, looking at both the kanga and also the second-hand clothing trade. Her grandmother in Philadelphia made a pineapple cake which was called the BTS Cake, i.e. Better than Sex. Paulette was to send it to me when she returned to the States, but sadly it never arrived. Perhaps it was censored. I can't claim that the Pineapple Fruit Cake recipe which follows is Better than Sex, but it is pretty good!

Pineapple fruit cake

400 grams (13 oz) fresh pineapple — blended but not too pureed

1 cup sugar

125 grams (4 oz) butter or margarine

2 $^{2}/_{3}$ cups of dried fruit

pinch of salt

1 teaspoon bicarbonate of soda

2 eggs, lightly beaten

1 cup plain flour

1 cup self-raising flour

1 teaspoon mixed spice

rum

This is a boiled fruit cake and any mixture of dried fruit may be used.

Place first five ingredients in a large saucepan. Bring to the boil over medium heat, stirring until butter is melted and ingredients are well mixed. Turn heat to low and cook for 15 minutes. Remove from heat and stir in the bicarbonate of soda. Cool — you can even leave it overnight. Add eggs, beating well. Sift dry ingredients and fold into the cake mixture. Place mixture in a greased 20 cm (8 in) cake tin, lined with two layers of greased brown paper and two layers of greased greaseproof paper. Bake in a slow oven 160°C (300°F) for 2 hours or until cooked when tested with a straw. Remove from oven and sprinkle with a spoonful or two of rum. Wrap cake in its tin in newspaper or a towel and stand on a cake cooler until it is cool. Remove from tin. This cake keeps very well — for two or three months in an airtight container.

It is a great deal better than the awful cake which Isaac, the gardener in Nairobi, made in my kitchen. He was attending classes at the local electricity authority on house-keeping, and along with ironing and cleaning the stove, he was taught a bit of cooking. He showed me his recipes one day and I asked if he'd like to practise something in my kitchen — the pineapple cake from last week's demonstration. 'O.K.', he said in his deep, deep voice.

I wrote down the ingredients I needed to buy, and managed to leave out the spice whose name I didn't recognise. Then it was impossible for me to remember, and the grocer and an African friend couldn't work out what it might be. So Isaac had to tell me again in his very deep voice that it was tangawesi powder that I needed. It sounded like something you would need for a voodoo spell, but it turned out to be just plain old ground ginger. It was a rather disgusting cake with a topping of crushed pineapple which one spread on the cake 10 minutes before it was finished.

My cake-making in Kenya wasn't a huge success partly due, I think, to the altitude. It is not only the oven temperature that is affected by the altitude. Cakes expand more in baking at higher altitudes and yeast dough rises much faster, although scones and biscuits and other types of baked goods weren't so drastically affected. When making cakes, each teaspoon of baking powder should be decreased by a quarter between 3000 and 7000 feet above sea level, each cup of sugar should be reduced by two or three tablespoons, and for each cup of liquid one should add three or four more tablespoons.

In Nairobi I kept a diary for more than two weeks, in pace with the creation of glacé or candied pineapple. Reading it two years on there are all sorts of domestic trivia which I certainly had forgotten and am not likely to remember again either. But there were funny little bits, like the very nice boy from whom I bought lots of my pineapples, who stuttered. Perhaps I should mention here, because he always sold me lovely fresh fruit, that the test for a ripe pineapple is simple. Smell it. You can tell if it is right. Then pull out a leaf and if it comes out the fruit is ready. They don't really ripen further in the family fruit-bowl but will slowly go off, and if kept too long may look perfectly all right on the outside but may be quite horrible when cut.

Glacé pineapple

1 ripe but firm pineapple
1 tablespoon salt
water
sugar
castor sugar

Day 1: Peel the fruit and remove eyes by cutting spirally downwards. Slice into 2 cm (1 in) rounds and remove core. Place slices in preserving pan and pour on four cups of water. Add the salt and bring to the boil. Simmer gently for 5 minutes and drain. Make a syrup of three and a half cups of sugar and four cups of water. Stir until sugar is dissolved and bring to the boil. Pour over the drained slices in the pan and boil for three minutes. Make sure that the fruit is covered by the syrup. You may have to weight them down with a plate. Leave for 24 hours.

Day 2: Drain the fruit but save the syrup. Add 55 grams (scant 2 oz) of sugar to the syrup. Bring to the boil and allow to cool slightly. Immerse the sliced fruit in the syrup and boil for three minutes. Leave to stand for 24 hours.

Day 3: Drain the fruit and save the syrup. Add a further 55 grams (scant 2 oz) of sugar to the syrup and boil the pineapple for three minutes. Leave to stand for 24 hours.

Day 4: Drain the fruit and save the syrup. Add a further 55 grams (scant 2 oz) of sugar to the syrup and boil the fruit for three minutes. Now you must leave it to stand for 36 hours.

Day 5: Drain the fruit and save the syrup. Add a further 55 grams (scant 2 oz) of sugar to the syrup and boil the fruit for 3 minutes. Again this must stand for 36 hours.

Day 6: Repeat as for Days 4 and 5 and leave for 36 hours. Technically with all these half days, we are now up to Day 8.

Day 8: Again repeat what you did on Day 4 and leave for 36 hours.

Day 9: Drain the fruit and save the syrup. Add 85 grams (scant 3 oz) of sugar this time, boil the fruit for 3 minutes and leave it for 48 hours.

Day 11: Repeat the procedure for Day 9 and leave for 48 hours.

Day 13: Repeat the procedure for Day 9 but add 115 grams (scant 4 oz) of sugar instead of 85 grams (3 oz). Boil the fruit and leave it now for 72 hours.

Day 16: Drain the fruit in a wire basket and save the syrup. Immerse the fruit in the basket in a pan of simmering water for three seconds. Drain and allow the fruit to dry. Cut the slices of pineapple into bite-size pieces and roll the fruit in the castor sugar to complete the drying process. Or if you prefer a glazed appearance, dissolve half a cup of sugar in one cup of hot water to make a new syrup and dip each piece of pineapple in this syrup. Dry the pineapple in a warm oven, turning it several times. Store in air-tight containers with waxed paper or cellophane between layers. The fruit will be good for more than three months if you can resist it. Then you can start all over again!

My diary tells me that during the time I did this, I went away for a weekend Museum excursion (it was during one of the 48 hour steepings of the pineapple slices) with Pat Duggan from the Australian High Commission. I was the navigator, and with the most pathetic map, I managed to take us nearly all the way to Garissa, a notoriously wild area, when we should have been on the road to Nyeri. We went hours out of our way which made us late for dinner at the Aberdares Country Club. I think the reason was really the mesmeric effect of the pineapple fields around Thika which is where I turned us in the wrong direction. Fortunately it didn't spoil my friendship with Pat, who even later gave me a pot of pineapple sage with little red flowers.

Another excursion I did from Nairobi was to lunch at Mrs Mitchell's. She was a doughty little person whose father had been one of the pioneers in the tea industry in Kenya. Her garden was beautiful and she gave the group of mostly American tourists a rousing talk on her family history. On the table at lunch was some very indifferent chutney so I seized the opportunity to ask her if I could supply her with mine, all the proceeds of which went to the New Life Home for orphaned and abandoned babies.

If you have a jar of indifferent chatni in your house a good way of using it up without having to really 'eat' it is in Cook Itself Chicken.

Cook itself chicken

1 good-sized roasting chicken

salt

450 grams (14 oz) crushed pineapple

1/2 cup fruit chatni

1 cup slivered almonds

1/4 cup prepared English mustard

Rub chicken with salt and lay skin-side up in a greased casserole dish. Combine the pineapple, chatni, almonds and mustard in a bowl and spoon over the chicken. Bake at 175°C (350°F) for 1 1/2 hours or until tender and well-glazed. Baste midway through baking and cover if it is getting too scorched on top. Serve with rice and salad and a sambal such as the pineapple sambal which follows.

Pineapple sambal or salsa

This accompaniment is a chameleon and can change its name to fit either a Mexican or an Indian meal.

$^1/_2$ a pineapple — peeled, cored and diced small

1 Lebanese cucumber — diced small

1 shallot — finely sliced

coriander or mint — a good handful of either, finely chopped

chilli sauce or ground chillis — a slurp or a pinch

Mix all together and serve in a bowl with a curry, a barbecue or what you will.

Of course none of the following chatni recipes are the slightest bit indifferent. I insist on spelling the concoction in the way closest to Hindi.

Pineapple chatni

1 medium pineapple — finely chopped

2 cups sultanas

2 tablespoons preserved ginger, finely chopped

3 cloves of garlic, crushed

1 medium onion — finely chopped

3 small red chillis — shredded

1 $^1/_2$ cups vinegar

3 cups brown sugar

1 tablespoon salt

$^1/_2$ cup of chopped nuts — I have used cashews or macadamias

Place all the ingredients, except the nuts, in a large preserving pan. Stir over a medium heat to dissolve the sugar and mix the ingredients well. Simmer until the fruit is tender. Stir occasionally to prevent sticking. Add the nuts and cook until mixture thickens. Remove from heat and pour into hot, sterile jars. Seal.

I like to shred the chillis rather than chop them finely so they are more identifiable and retrievable for those who want to avoid them. Of course their flavour permeates the mixture but a shred can be removed if desired. Makes about five cups of chatni.

Pineapple & orange chatni

2 pineapples — chopped fairly finely — the prepared fruit should weigh approximately 2 kg (4 lb)

3 oranges

2 cups vinegar

1 cup brown sugar

2 small onions, chopped

1 clove garlic, crushed

1 teaspoon salt

1 teaspoon ground allspice

1/2 teaspoon chilli powder

1 stick cinnamon

1 cup chopped, pitted dates

1/2 cup sultanas

Peel one orange carefully and thinly and cut the peel into fine slivers. Remove peel and pith of all the oranges and separate into segments. Check for pips and remove them. Mix together the vinegar, sugar, onions, garlic, salt, spices and dried fruit in a preserving pan. Stir as you bring it to the boil. Add the slivers of orange peel and simmer for half an hour. Add the pineapple and simmer uncovered for an hour or until thickish. Add the segments of orange, stir well and simmer for another 15 minutes. Pour into hot sterile jars and cover.

Pineapple & ginger chatni

This was originally an apple chatni which is equally successful using either quinces or pineapple.

3 kg (6 lb) pineapple

1 kg (2 lb) sultanas

375 grams (12 oz) preserved ginger

1600 ml (3 pints) vinegar

1 3/4 kg (3 1/2 lb) sugar

30 grams (1 oz) salt

1 teaspoon ground allspice

Peel, core and chop the pineapple into small dice. Chop ginger. Mix vinegar, sugar, salt and allspice in a large saucepan. Bring to the boil and simmer 10 minutes before adding the fruit. Cook until the mixture becomes thick and dark, stirring now and then and checking that it doesn't stick to the bottom of the pan. Pour into sterile jars. Cool and seal.

I saw a strange manifestation of the pineapple close to Kota, another important centre for the block-printing of cloth in Rajasthan. It was at Bundi, like Kota a former princely state, and it was my birthday. I was all by myself and the day had been spent looking at the beautiful Bundi murals in the Palace and a scramble up the slippery stone path to the fort on the top of the hill. My bargain for the day had been new heels on my sandals, complete with about 35 nails, for

5 rupees which is the equivalent of about 20 cents. The cobbler had even offered me a cup of tea!

The sun was sinking and the heat of the day was ebbing so I went for a walk back into the bazaar. An itinerant toy-seller was walking along the street in front of me carrying a stick surmounted by his wares. They were simple things made of cardboard, a 'person' on a stick and strings attached that when pulled made the arms wave and the legs dance. Most of the 'people' were film-stars from Hindi movies, but the real star on the top of the pole was a fair and chubby baby in a pink turban, clutching a most unsuitable plaything — a pineapple! It was one of those moments when it is a shame to travel alone. I did laugh out loud, but I really wanted to exclaim to a friend. However, I bought one as my birthday present to myself.

It cost 10 rupees, being the superior version. The film stars were priced at only 5 rupees. Now I come to look at it, I find that there is a change in character for this baby. He has been accorded extra limbs in order to become articulated and because of this he has taken on an almost divine identity. Perhaps he will become something like Lord Krishna who is worshipped by many in the form of a rather mischievous baby. With his four arms who knows what this pineapple-embracing god will be able to do.

Toyseller in Bundi, India

The pineapple baby god — a simple toy for children bought at Bundi in India. In front are china teapots from Thailand and jarlets from Mexico

Ananas achar–Pineappple oil pickle

2 large pineapples
20 red chillis (optional)
salt
2 1/2 cups vinegar
2 teaspoons fenugreek seeds
1/2 cup peeled cloves of garlic
1/2 cup chopped green ginger
1/2 cup ground cummin
1/2 cup ground coriander (A cup of dhanajeera powder can be substituted for the cummin and coriander combined)
2 1/2 cups vegetable oil such as safflower (never blended)
2 tablespoons black mustard seeds
2 tablespoons curry leaves
1 1/2 tablespoons fennel seeds
1 tablespoon ground turmeric
1 tablespoon chilli powder
1/2 cup sugar
2 1/2 tablespoons salt

The night before you plan to make the pickle, soak the fenugreek seeds in a small quantity of vinegar. Next day, place the fenugreek and soaking liquid with the garlic, ginger, cummin, coriander and the rest of the vinegar into the blender. Blend at high speed to make a puree. If you are using chillis, split them in halves, almost up to the stem end. Peel and core the pineapples and chop into dice. Put into a colander over a saucepan to catch the juice and squeeze out as much juice as possible. While you continue with the recipe you can sip at the retained juice. Thus fortified, heat the oil in a large enamel or stainless steel saucepan until very hot. Add mustard seeds, curry leaves, fennel, turmeric and chilli powder and then the blended mixture. Fry, stirring constantly for 2 minutes. Then add the chopped pineapple, chillis and sugar. Bring to the boil and then reduce heat and simmer uncovered for about 30 minutes or until the oil rises to the top and the spices smell cooked. Add salt to taste, cool and bottle in sterilised jars. Oil should cover the top of the pickle. If there isn't enough, heat more oil to top up the jars. On the other hand, if there is too much, leave it in the jars but when you go to make a meat or vegetable curry, remember it is there and use this excess oil to start it off. It is utterly delicious. Remember always to use a dry spoon when taking any oil pickle from the jar.

I used to give jars of this particular pickle to a dear old man at the shop where I went in Nairobi to have my films developed. He was a devotee of Sai Baba and once, when he heard that I had been sick, he gave me a film spool container full of amrit, or sacred nectar, that had come from the Sai Baba Ashram in south India. He said it would make me better, and sure enough it did. It is a matter of faith.

He was in his 70s and had been born in Kenya but his native place was in Jamnagar in Saurashtra, the area where the Portuguese developed their liking for such Indian textiles as the palampore and the place whence the kanga sprang. (That sounds like something to do with Australia's favourite animal!) He went to India for his sister's wedding in 1942 and sailed back in a dhow. Because it was wartime, fewer vessels were making the trip and his dhow, which should have had a complement of 100 passengers and crew, sailed with 250 on board. Instead of taking 15 days, the voyage took 28 days. He said it wasn't a lot of fun, with dry rations and one cup of water a day, and of course no bathing.

He told me all sorts of stories and one day when I arrived to collect my latest film he became quite convulsed with laughter. This was because there were quite a few photographs of me stark naked, but not revealing a thing, in the askari house I had had made which we kept in the house. He was far too polite to say anything — just roared with laughter.

Askari houses are little sentry boxes that unfortunately are a necessity at one's gate in Nairobi. A poor askari, or watchman, sits there all night — all too often falling asleep — to guard the household from robbers. I thought that one in splendid colours would really improve our entrance hall, with its ever-so genteel, pale pink flowered wallpaper. So I commissioned one to be made by one of the *ad hoc* carpenters on the Lower Kabete Road. It was to be dark blue on the outside and a dark bright red within, and was a great success. People coming to the house for the first time didn't quite know what to say. 'Oh, oh...inside?...Oh, you've got one of these inside...oh...what fun.' It *was* fun and for some reason my husband thought I should pose naked inside it. Why, I do not know, and I resisted strongly. It just so happened that I received from America a card with those well-known photographs by Robert Mapplethorpe of a muscular young man posing naked on a pedestal in a very modest manner. So I thought to myself, well, if he can do it I can too and so I did.

On usual days I used to have flowers sitting on the seat of the askari house and on the back wall a self-portrait by Mazola, a good Kenyan artist, who became a friend. On the floor was a family of wooden guinea fowls with eggs in a basket nest. Sadly, it was too big to fit into our container to bring home but it was bought by an American and it is now in Los Angeles. My husband made me an Australian version which was unveiled by Pat Duggan, whom I had misdirected through the pineapple fields of Thika. She cut the ribbon with my foot-rot clippers while wearing a pair of marvellous rubber gloves adorned with rubies and leopard-skin. It all seemed highly appropriate at the time.

When I started making such prodigious quantities of preserves in Nairobi, I sold them at first at the Australia New Zealand Women's Group meetings and gave them the proceeds, but then I was introduced to the New Life Home. What a happy, lovely place it was. It was run by Mary and Clive Beckenham who, having lived in Nairobi for a few years, had become aware of the plight of abandoned babies. Half of these poor little things were HIV positive and nobody wanted them. They had been found in the most horrible places — in pit latrines, in rubbish bins, in boxes, and were all in a very sorry condition when they arrived to start their new life. For this is what most of them did. A very few extremely sick little ones did die but the great majority survived and thrived to show vivid contrasts to the photographs taken when they arrived at the home at Loresho.

The majority were baby boys, as in Kenya the women do all the work, and, unlike many societies, the boy babies are the unwanted. The baby girls grow up to continue to do all the work.

The most miraculous thing of all, however, was not the way the thin, frail little infants turned into round, healthy little people. As I write this, I have just received a letter from the home which says that of the 34 infants who had been HIV positive, there have only been two who, when retested after months of love and care and vitamins and a proper diet, have not sero-converted. While I lived in Nairobi a scientific journal had published an article recording similar findings, but the New Life Home had known about it for ages.

When I began to explore Nairobi, I was frequently aware of people looking at me in a certain way. It had also happened a few years before, in Australia, so I knew what it was. I would feel a stare when I was somewhere like a supermarket and think, oh no, the mistaken identity again. I happen to look rather like a character in the television show *'Neighbours'*. Unfortunately, she was a ghastly character so it wasn't as funny as it could have been — she was Aunt Hilary. She had been written out ages before in Australia, but the programmes being run in Nairobi at the time were old, so I had to go through it all over again. I would sense that certain stare and then hear comments sometimes in English or sometimes just the words 'Aunt Hilary' coming through in a Swahili conversation.

Strangely enough, a 'Neighbours' crew arrived in Nairobi to film a story-line there. The programme in Kenya was sponsored by a company which sold washing powder. The New Life Home, with 12 little babies in nappies, used a lot of washing powder so I had what I thought was a brilliant idea. Why not use me as Aunt Hilary in an advertisement for the washing powder and give the money to New Life? So I wrote a letter, took one copy down to the Norfolk Hotel where the crew was staying, and another to the soap powder company. I drew a complete and utter blank. Thus my career as an Aunt Hilary lookalike just didn't ever get off the ground. I have to admit I wasn't sorry.

The Kenyan fruit vendor's stall where I usually bought my pineapples, was situated just down the road from our house, a matter of about a quarter of a mile. Sometimes however, it could take up to 20 minutes to get there. Our house was next but one to part of the campus of Nairobi University. It was the veterinary school to be precise and there was a freezer — to keep dead animals — and a furnace for their disposal. It had unfortunately become an unofficial funeral parlour — for humans — and nearly every day we had to negotiate great throngs of people and cars in a narrow little lane designed for no traffic at all. One could hardly beep the horn at mourners so we developed patience, but we still would have preferred not to have a pile of tin coffins on the rubbish tip as a view.

Some serious pineapple sleuthing in India took me to a graveyard. This was to the South Park Street Cemetery in Calcutta. The first time I visited the cemetery it was in a state of great dilapidation. Many of the splendid monuments had been vandalised and the prolific tropical vegetation had grown up between, and all over, great pieces of masonry. Ravens cawed in the trees and lizards scuttled underfoot. People had made their homes in the houses of the dead, some of which were like Palladian rotundas, others like Grecian temples, and teams of mean-looking dogs were on patrol.

I revisited what is known as the Great Cemetery at the end of 1995, and found still the frightening canine bands but a great change in the landscape. A restoration campaign was begun in 1978 and continues. Tombs have been restored and missing plaques have been replaced. Paths have been relaid, drains repaired, a running water supply laid on, retaining walls built and the imposing front entrance has been reconstructed. Tonnes of vegetation and rubbish have been removed and the central area has been replanted with bougainvillea, poinsettia, oleander and other fast-growing colourful shrubs. Funds have come from firms and individuals in both Calcutta and the United Kingdom.

The superintendent, who had Spanish ancestors, was a very obliging man called Raphe Arthur d'Riego, known as John. He took me to see the grave of Rose Whitworth Aylmer who departed this life on 2 March 1800, aged 20 years. According to William Hickey, one of the great diarists of the world, she died from eating too many pineapples, or in his words, after 'a most severe bowel complaint brought on entirely by indulging too much with that mischievous and dangerous fruit, the pineapple'.

I had thought I would read on her tombstone of her sad fate by ingestion. 'No, no,' said Raphe Arthur, known as John, 'can't you see? The whole tomb is like a pineapple dressed for the table.' I suppose with a little stretch of the imagination it is. Among the huge pillars and pyramids, Rose Aylmer's memorial is different. It is a tapering, twisting column garlanded two-thirds of the way down with a beribboned swag of leaves. A little too tapered to be a pineapple, but if one allows a certain artistic licence, yes.

Rose Aylmer's grave, Calcutta, India

Rose had come out to Calcutta at the age of seventeen, to join her aunt, Lady Russell, whose husband, Sir Henry Russell, gave his name to Russell Street, not far from the cemetery. Whether she wanted to come or not we do not know, but she left a broken heart behind her in Wales — that of the poet, Walter Savage Landor. They had fallen in love at the Swansea Circulating Library. After her death, which the cemetery records say was due to cholera, Landor wrote an ode which became one of the favourite poems of Charles Lamb. It was added to her tomb in 1910, subsequently vandalised, but now restored:

Ah, what avails the sceptred race!
Ah, what the form divine!
What every virtue, every grace!
Rose Aylmer, all were thine.
Rose Aylmer, whom these wakeful eyes
May weep, but never see,
A night of memories and of sighs
I consecrate to thee.

Oh, poor Rose Aylmer.

William Hickey, the chronicler of Calcutta, has written much to illuminate the history of the city that was at this time the second city of the Empire. He must have had the keenest eye and quite obviously no-one was safe from his recordings. He was circumspect, however, or perhaps nobody suspected his pen catalogued their every activity. His published *Memoirs*, in four volumes, cover the period 1775-1809. He was a lawyer and therefore a member, along with the doctors, engineers and chaplains, of the small middle class of the British settlements.

At the top of the social structure were the soldiers and officials of the East India Company and the servants of the King. According to Percival Spear in *The Nabobs*, the army was a stronghold of a vigorous philistinism. Only among the civil service was there any interest in culture. It existed too among the small, professional, middle class, whose tastes and styles were copied from the officials. Lawyers were the only legitimate members of this professional middle class as doctors and engineers were generally Company officers with military rank. Hickey was a prosperous, if rather profligate lawyer and the name of his firm was still to be found on documents as late as the 1970s. He was a keen observer of all that went on around him and wrote not only of personalities, legal intrigues and politics but of commonplace objects such as hookahs and the changes in smoking habits, of punkahs or swinging fans, of his zenana or harem on the banks of the Hughli [sic]. He wrote too of servants — he had 63 by the time he left India, of food and 'superabundant potations of champagne and madeira', of parties and the colourful dress which, though quite *de rigeur* in Calcutta, was very *outre* back in England. And of course, he wrote of the deaths and funerals that punctuated existence in Calcutta.

Rose Aylmer died from eating too many pineapples; a general, in 1812 died from eating too many radishes; a beautiful young woman called Emily Warren, painted many times by Sir Joshua Reynolds, died on the banks of the Ganges from drinking too much milk. Drunkenness, too, was a regular cause of death. English habits of eating and drinking had not recognised the huge difference in climate and the normal diet was entirely unsuitable for the Indian summer. A massive meal was taken in the very heat of the day with soup, roast fowl, mutton pie, roast lamb, curry and rice, puddings, tarts, cheese and lashings of alchohol. No wonder people fell down dead afterwards.

In Australia, at the Canberra Word Festival in March 1997, one of the panel discussions was on the subject of 'Bliss: the delights of Mortality'. It was a most wonderful discussion with Robert Dessaix, Carmel Bird and John Scott as the speakers. Robert Dessaix, to whose writings I am totally devoted, was wonderful and John Scott was great, but Carmel Bird was entirely captivating with her talk about old graveyards in Tasmania, populated by phallic headstones, and about having mild sex with nice boys in a Melbourne joss-house and...did she really 'do it in the back of an old hearse'?

She said that man is a tomb-building animal and that perhaps someone should establish a graveyard for Australian novelists called the Valley of the Beautiful Liars. She had visited the Forest Lawn cemetery in California and had found it like a theme park of modern Western death.

I have always enjoyed old graveyards. St Matthew's at Windsor, west of Sydney, which was in my father's federal electorate when I was a child, was my first introduction. But the necropoli of the British Raj in India really are the best.

In Ootacamund in south India I saw the tomb of the Hon. Henry Handcock, Aide-de-Camp to the Governor of Madras, who was killed by a tiger in 1858. Search as I might on several occasions in Peshawar in the North West Frontier Province of what is now Pakistan, I could not find the resting place of Captain Ernest Bloomfield. His marker is said to be inscribed thus: 'Accidentally shot by his orderly, March 2nd 1879, Well done, good and faithful servant.' There is, however, definitely a monument to the poor souls who died in an explosion at the soda-water factory.

I gave myself a terrible fright in the Bankipore Cemetery in Patna in August 1995. It is a very old cemetery with graves in the same eighteenth century style as those of the Park Street Cemetery in Calcutta. When I was there it was during the late monsoon and the vegetation was dense and choking all the paths. After a while, and hearing ominous rustlings in the undergrowth, I thought to myself, 'I had better get out of here, this is just the place for a snake'. That thought had hardly passed through my head when I felt the most furious bite on my foot. 'Too late,' I thought, 'oh well, I'm in a cemetery'. And I leapt onto a tomb to find only a large and angry ant attached to part of my foot not protected by sandal.

I had had quite a morning, having had to abandon a cycle rickshaw and walk through a barrier of shouting protesters who were stopping all vehicular traffic. I had negotiated a second rickshaw and finally arrived, after an hour's pedalling, at Quila House, a stately mansion on the banks of the Ganges. It was the private museum owned by the Jalan family. I had tried to telephone from my hotel without success but I was met at the door by the family. It was obvious that they were just about to go out but with great grace they insisted on showing me through their beautiful home.

The collection had been created by family members for generations. They have a wonderful textile collection, but it was the wrong time to view it in the humidity of the monsoon. There is an important collection of Chinese celadon platters, Louis Quatorze furniture and even Napoleon's bed!

I was given a glass of water and then the son, Aditya, showed me through the rooms. Then I was given *nimbu pani* or lemon squash, in a silver cup and a delicious milk sweet decorated with *chandni warak* or edible silver leaf served on a silver plate. After this they gave me a lift to the Bankipore Cemetery which I imagine they thought was rather an odd destination.

From here I tried my level best to direct the next rickshaw-wallah to the old opium factory, but without success and so I had a long walk which took me to the quarter where the Law Courts are situated, again on the banks of the Ganges, which at this time of the year was huge and swollen and brown. Vendors sitting in the sun at the roadside sold documents and forms to plaintiffs and defendants, scribes and typists sat at ancient typewriters on battered tables under banyan trees, and harrassed-looking lawyers in black coats and grubby jabots hurried about. Nearby was a nursery school in the grounds of the old Anglican Church. Little children were leaving the morning session in conveyances like chicken coops — funny wire-cages atop cycle-rickshaws.

I was in Patna for the first time in 25 years in order to visit Madhubani, further north in the state of Bihar. I wanted to see for myself the vibrant paintings which the women of this area paint on the walls of their houses. My friends in Delhi had said, 'Oh you can't go there by yourself, there are dacoits (bandits)'. The people in the hotel in Patna said, 'Oh you can't go there by yourself, there are floods'. But I went, and had a glorious day, without encountering even one *dacoit*. We drove past occasional palaces and floodwaters were everywhere except on the road. Mongooses, or is it mongeese, scuttled away into the undergrowth, avoiding the car. Children carried black umbrellas as they rode on the backs of buffalo, and a crowded steam-train chugged past us with dozens of passengers sitting on the roof .

I met a wonderful old lady, Shrimati Mahasundari Devi, a National Award Winner, who showed me her work and that of other members of her family. I was taken all over the home of the extended family and saw the colourful paintings of gods and goddesses, animals and flowers. We had tea and she picked flowers from the garden to show me the colours that she used. A little blue convolvulus and a blood-red hibiscus were just two. Her paintings, transferred from wall to paper, are often in just red or just black and show the finest lines which she paints with a limp thread taken from the frayed edge of a piece of cotton cloth. It is quite unimaginable.

Two Monsoons is the title given by Theon Wilkinson to his book, a saga of European mortality in India. Two monsoons, was, at least in the eighteenth century — the average survival period. Well, if the Californian cemeteries are theme parks of the modern western way of death, I suppose the old graveyards of the British in India are Raj theme parks. In comparison, there may be little atmosphere in modern cemeteries because death usually isn't quite so graphic and dirty as it was then.

One person who successfully managed to last many monsoons in India was General Count Benoit de Boigne. He was the mercenary general who commanded the great army of the Mahratha ruler, Madhaji Sindia of Gwalior. He was born in 1751 in Chambery in France, the son of a trader in hides. The addition of 'de' was bogus and it is clear that, from the start, Benoit la Borgne — he changed that spelling too — had great plans for self-elevation.

Stories say that he had been a lover of Catherine the Great, who, tiring of him, had sent him off to the Turkish war. He was captured and sold as a slave in Istanbul but continued on adventuring and meeting the right people at moments of crisis with an uncanny regularity. He arrived in Madras in January 1778 and, after a series of false starts and misadventures, he appears in Madhaji's camp, besieging the fortress of Gwalior in August 1783.

He had been assisted thus far by characters who loom large on the stage of Indian colonial history — Warren Hastings and Claude Martine among them. De Boigne seems to have captivated everyone he met. He spoke English with an Irish accent, having been a member of the famous Irish Brigade of the French army, and perhaps it was this touch of the blarney, or the gift of the gab, that helped him on his way.

When he met Madhaji he was penniless, all his baggage having been stolen, probably by Mahratha thieves in Madhaji's service. However, not only was he dignified and handsome, he was shrewd and diplomatic. And he knew then, as one still finds today, that in Asia it doesn't do to lose one's temper and cause other people to lose face. Instead of just retrieving his lost possessions, he won himself quite a job. He remained in the service of Madhaji, and his successor Daulat Rao Sindia, for 11 years. He raised what came to be called the Army of Hindustan, led it to a series of victorious battles and became Sindia's Governor of Hindustan, the last remaining province of the Moghul Empire. What's more, he lived to tell the tale. He made his fortune, returned to France where he received high honours from Louis XVIII and Victor Emmanuel who in 1822 created him Comte de Boigne. He lived until 1830.

Whether on manoeuvres or at home in his cantonment at Aligarh, de Boigne lived in great style with his beautiful, well-born Muslim wife. She may have been the daughter of a Persian cavalry officer or she may have been a princess, but she wasn't present at a feast described by a Mr Twining in 1799. Dinner was then served at four, 'much in the Indian style; pilaws and curries, variously prepared, in abundance; fish, poultry and kid. The dishes were spread over a large table fixed in the middle of the hall, and were, in fact, a banquet for a dozen persons, although there was no one to partake of it but the General and myself.'

Perhaps when the General was just in need of a bit of a snack there were fewer courses and perhaps that which follows on the next page is what he ate. This dish is said to have been invented by him. The marinade, a little Gallic touch perhaps, may have been to disguise not only the toughness of the beef but any suggestion that it might have been past its best.

Pineapple beef or Boeuf à la Benoit de Boigne

500 grams (1 lb) sirloin steak — cut into rounds to match the size of the pineapple rounds

4 cloves of garlic, crushed

salt

$^1/_2$ teaspoon chilli powder

2 tablespoons of either Worcester or soy sauce — I think soy is by far the nicer

1 tablespoon oil

1 small pineapple, peeled and cut into rounds

a little more oil

Mix salt, chilli powder, sauce and oil to form the marinade. Rub the beef rounds with garlic and place in the bowl with the marinade. Cover and leave for a few hours or overnight in the fridge. Heat the griller and grill the steaks and while you do this sauté the pineapple slices quickly in a little oil — all the better if the oil is flavoured — perhaps you can pour a little off the top of your jar of Ananas Achar (see page 58). Top the steak with the fried pineapple and garnish with coriander.

Serve with potato curry and a green salad. Serves three.

It is impossible to imagine Indian food without the addition of chillis, fresh, dried or pickled but they have been in India only as long as the pineapple. Before the arrival of the Europeans in India, the only 'hot' ingredient in food was pepper, although there could have been an indigenous chilli in Kashmir. The Portuguese established trading factories in Cochin and Calicut in 1501 and after this date the chilli began to be cultivated. As well as the wonderful pineapple, the Portuguese also brought other important food crops including the papaya, the cashew nut and the tomato.

The inventories of deceased estates of citizens who had lived in India at the beginning of the eighteenth century are interesting lists. There is often a box of pipes, furniture and a 'pallaqueen' — a palanquin was a mode of transport about the town. There could be 'a pleasure boat, a silver coffee pot, 17 old snuff boxes, a blunderbuss, a slave boy, a bayonet and two meddals [sic], a cask of tobacco, a long old wig, gowns, petticoats and fanns [sic], a banyan coat, silver breeches buckles, ten banyan shirts, eight night capps [sic], two teapoys, one garden bungalow, eight silk wastcoats [sic], a tonguescraper and three Palampores'.

The word palampore comes from a corruption of palangposh, meaning bedcover. The designs on these large, mordant-painted and printed cotton cloths were varied, but the most enduring and that which had the most lasting effect on the textile traditions of countries other than India, was the tree of life pattern. In the seventeenth and eighteenth centuries, huge quantities of these cloths were exported to Europe. The

Portuguese, in particular, liked these 'pintadoes' or painted cloths, which they commissioned in their colonies on the west coast of India. Other cheaper printed cloths were produced in Diu for trade with East African coastal settlements, so we are again reminded of the kanga and its history. These more utilitarian cloths were used and discarded but old palampores exist in museums all over the world.

The traditional palampore had eclectic design influences from China and Iran. It generally shows a central tree growing from a rocky knoll, surrounded by exotic flowers, fruits, birds and animals such as tigers and peacocks. It was also traded east to mainland and archipelago south-east Asia. Like other Indian textiles such as the patola — the double-ikat cloths of Gujarat — the palampore was regarded as extremely precious and achieved the status of a sacred heirloom. The tree form appealed to both Islamic and non-Islamic people and was reinterpreted and enfolded into Hindu-Buddhist symbolism as another form of the cosmic mountain. Torajan ma'a, or sacred heirlooms, from Sulawesi, and Cirebon batiks from the north coast of Java, are just two textile forms which show influence of the beautiful palampore.

Much further afield, the palampore had a certain influence on a quilter who lived in Philadelphia at the turn of the nineteenth century. The Rachel de Puy Quilt, which is dated 1805, is a most beautiful white on white quilt sewn in the Marseilles quilting technique. In this form of quilting a small running stitch is used to outline the motifs and then the quilter carefully separates the layers and inserts cotton cording or batting to give more dimension to the areas she wishes to enhance. The palampore tree of life occupies the centre of this quilt and is surrounded by shepherd and shepherdess, flocks of sheep, twining flowers and, symmetrically arranged in the upper right and left, either side of the top of the tree, are two pineapples.

Pineapples, as we have seen, were adopted by the Americans as a symbol of hospitality and are depicted in all manner of the applied arts including quilting. In this particular quilt we can see a few lovely historical links. The Portugese introduced the fruit to India and it was most likely here that some travellers came into contact with it for the first time. Guests to European tables in India certainly dined on them and perhaps they may have even slept under them — as printed images on the palampore bed-covers and canopies in the guest-rooms.

Some foods from the New World were introduced to the United States of America from Europe, even though it was part of the New World itself — a very circuitous route. Perhaps the creator of the Rachel de Puy Quilt had come across pineapples through those hospitable sea-captains coming back from the West Indies and using them as an open house sign at the gate, but she certainly received inspiration from the Indian cloth.

The symbol of hospitality had another place in the history of American patchwork quilts. The pineapple became a popular pattern — a variation on the Log Cabin — after 1875. Another name for it is Courthouse Steps.

A carnation and pineapple design applique quilt has a connection to Garrard County in Kentucky. A beautiful quilt with 'real' pineapples, so different from the prickly pine-cone shapes of the log cabin derivative, is to be found in the Helen F. Spencer Museum of Art at the University of Kansas, where 20 pineapples, like glorious cartouches, are lined across the quilt.

The American patchwork quilt has its roots in many cultures. Patchwork quilts were rare before 1800 and were much more likely to have been owned by the rich, and not the poor, contrary to the general view of frugal bits and pieces stitched painstakingly together. In truth, patchwork demands a certain minimum level of affluence and technology because a diversity of fabric is required. Cloth, before the development of factories in the industrial revolution, was scarce, and people didn't have many clothes. Many of these garments too were made from economical square-cut patterns which used cloth from selvage to selvage, leaving no scraps at all.

Asia is generally regarded as the cradle of patchwork and applique but it is found in Africa and all over the world. It is the economical utilisation of partially worn-out clothing. It is important in Islam and in Buddhism the pieced robes of monks are made from patches.

In Bangladesh needleworked quilts are called *nakshi kanthas*. I had read about them and had seen beautiful old examples in the Folk Museum in New Delhi. I tried without success to find them on two visits to Bihar, which is one state in India where they were traditionally made, the other being West Bengal.

The word kantha comes from the Sanskrit for rags and the traditional manufacture is to use cast-off saris or dhotis. A simple running stitch was used to quilt three or four layers of cloth and colourful pictorial designs were filled in with satin and stem stitch using thread drawn from the coloured borders of the original garments.

The creation of decorated kanthas for home consumption in West Bengal has largely died out, although it exists still in Bihar. In the independent state of Bangladesh the tradition is not only alive and well, but has taken on a new vitality.

The name nakshi kantha comes really from a poem written in 1939 by Jasim Uddin which in translation is titled 'The Field of the Embroidered Quilt'. He was perhaps the first person to recognize the significance of the kantha and its association with the lives of the women of rural Bengal. He succeeded so well in immortalizing it that the appellation nakshi, meaning figured or decorated, prefaces the word kantha whenever Bangladeshis talk of the quilts. It does in fact distinguish the embroidered and often very sophisticated quilts from the humble, unadorned ones that can be seen in villages.

The making of these quilts was becoming a lost art until after the liberation of Bangladesh and the revival is perhaps linked to two factors: a sense of national difference which inspired the resurgence of indigenous crafts, and the need to provide for women left destitute after the war. A start was made in 1972, but the real revival did not get underway for another ten years with the help of several non-government organisations.

Another important landmark was the opening in 1981 of a new five-star hotel in Dhaka – the Sonargaon – where large and very beautiful kanthas were a focus of the decor, showing pride in indigenous crafts and inspiring visitors to acquire one for themselves, although these are now made from new cloth and not from cast-off saris. I stayed a couple of nights with my daughter, Emily, in this very nice hotel. We both bought kanthas and one of mine had some Bengali words embroidered on it. It was like drawing teeth to try and get the young men in the shop to translate it. They obviously thought it far too personal and suggestive for it turned out to be an avowal of love by the embroiderer for her husband.

After this first purchase I bought them at many shops in Dhaka. There is an enormous range of size and pictorial design. Village life and nature are observed, Islamic geometric designs are incorporated into some versions, and others show Hindu motifs such as lotus mandalas and ceremonial cars.

The kanthas are divided into four major styles according to the type of stitch used. The ones made with running-stitch are called lik. Of these the lik phul or flower stitch is a popular pattern, where parallel rows of stitches form a flower-like shape. This stitch is commonly used on red cloth at Chapai Nawabganj and on white at Jessore, both in the west of the country, and another name for it is anarasi—or pineapple. I discovered this pattern in Dhaka at the very time when my husband was collecting for me in Laos a length of indigo ikat silk woven by my great friend, the weaver, Kongthong Nanthavongdouangsy. The pattern of the ikat is pineapple and there is some resemblance to the Bangladeshi stitch design.

An old patchwork quilt of the pineapple pattern made of velvet and silk

Pineapple patchwork pattern

Pineapple embroidery stitch from Bangladesh

In Dhaka I also bought some pretty little wooden spoons with painted handles in the shape of pineapples. Unfortunately they taste quite musty and disgusting and even transfer this ancient mouldy flavour of 102 monsoons to the chatnis they might be spooning, so they are languishing in the cutlery drawer.

The other coincidence in Dhaka was that while there I received a letter from my husband with Kongthong's recipe for a green dye for textiles using the leaves of the pineapple. That day I happened to visit Sayyada Ruby Ghuznavi in her office and workshop. She has done a great deal of research into the history and development of traditional vegetable dyes in Bangladesh and actively promotes the use of these dyes instead of synthetics. I bought a pale blue silk shawl dyed with indigo and a deep red scarf dyed with manjit. I told her of Kongthong who has done similar work in Laos and of how I'd just received the pineapple recipe that morning. 'What is the mordant?' she asked. 'A few rusty nails', I replied, and she promptly asked if she might have the recipe. So there now is a bit of cross-fertilizing of vegetable dyes. Here are the instructions for green dye.

Green dye for silk

1. Take 1 kg (2 lb) of fresh leaves.
2. Beat them into a pulp—takes about half an hour.
3. Mix 5 grams (one teaspoon) lime in 5 lt (20 cups) of cold water (if no lime use half a dozen rusty nails).
4. Add the pulp to the lime and water and stir.
5. Remove the pulp (and nails).
6. Soak the $^{1}/_{2}$ kg (1 lb) of silk in the dye for half an hour (raw silk must have been previously boiled).
7. Remove silk and dry in the shade.
8. Wash the silk in cool water.

If the season has been very dry then these leaves will produce a much darker dye.

As I explained in my preface, Kongthong also sent me the Lao riddle. It always amazes me how some people can remember stories and jokes. I was talking about pineapples to Richard Begbie, an old friend, who writes a regular column for the *Canberra Times* and he told me something he had heard 30 years ago. The story took place in 1943 in war-time London when a Queensland serviceman on leave, feeling desperately homesick, set out to find a pineapple. After hours of fruitless search (oh, I just couldn't resist that) he saw one arranged in pride of place in the centre of a greengrocer's window. He went in and asked the price. It was 17 shillings and sixpence, which just about floored him as he was used at home to them being threepence a dozen. But he

Indigo ikat silk from Laos showing the traditional stylised pineapple design

A silk pha biang or shoulder-cloth from Laos dyed green with pineapple leaves. Eucomis or pineapple lilies fill Thai pineapple patterned china vases. In front are two enamel pineapple perfume containers

Crocheted doiley of pineapple design

had to have it and handed over a pound note. He received his parcel and turned to walk out and the honest fruiterer said, 'Hold on, you've forgotten your change.' 'Don't worry,' said the Australian, 'I trod on a grape as I came in.'

The fact that this serviceman found a pineapple may have meant that it was close to Christmas. A notice was put in The *Times* — in the normal peacetime season of peace—even as late as 1969, that at Harrods 'the pineapples have arrived for Christmas'.

I always thought it was most peculiar and downright conniving deprivation on the part of

Scrooge-like parents who gave little children in England an orange in their Christmas stockings, and amazing that the children were thrilled to bits to find the oranges. Mean old Santa I thought. Boring old oranges. For those of us spoiled by temperate climates all the year around and especially today when we buy, as a matter of course, lemons and garlic from the United States of America and, if feeling profligate, mangoes from Mexico out-of-season, it is too easy to forget the cycle of nature.

Perhaps it would have been a nice touch to have tied in the Harrods advertisement with the latest production by the Royal Ballet of 'Pineapple Poll'. It is one of the two most enduringly popular English ballets and was produced for the first time in 1951 with the choreography of John Cranko. Ninette de Valois, the legendary director of the ballet, wanted to include in the repertory 'works encouraging a strictly National flavour'. The music was by Sir Arthur Sullivan, unaccompanied on this occasion by Mr Gilbert although the ballet was based on Gilbert's 'The Bumboat Woman's Story' and set in Portsmouth. The first pert little Poll was Elaine Fifield and sets for the first production were designed by Osbert Lancaster. There is an important Australian element as the music was arranged by Sir Charles Mackerras who is the brother-in-law of my great school-friend and literary mentor, Lindsay.

I don't really like serving food in pineapple shells, particularly hot savoury food. A very famous wedding breakfast began with a hollowed out pineapple half filled with something which I am sure must have been delicious. It is hard to tell from the photograph what it is that the beautiful young bride and groom are about to eat. The food must be cold as the table is set and laid with the entree before the meal begins. The scene is a garden and there must be many guests, but the groom is so hungry that he is about to set to on the pineapple even before his bride is seated, let alone the rest of the bridal party. The bride and groom were Jacqueline and John Kennedy. I guess there is something symbolic in starting one's married life with the American icon of hospitality. We can just imagine what the rest of the dish was — I hope it wasn't just chopped mint and maraschino cherries. Perhaps it was chicken salad and pineapple.

Chicken salad with pineapple

1 1/2 cups Lebanese cucumber, chopped into small dice

1 1/2 cups pineapple, chopped into small dice

1 1/2 cups steamed chicken, chopped into small dice

3 tablespoons mayonnaise

1 1/2 teaspoons nam pla, Thai fish sauce — or to taste

1 teaspoon nam prick, Thai sweet chilli sauce — or to taste

chopped coriander

Mix all together and serve garnished with coriander. If you are going to double quantities and serve in a large bowl lined with lettuce, a central 'flower' with petals of avocado and tomato looks good. Don't mix too far ahead as the pineapple will lose liquid and the mixture will be a little too wet. Serves four.

I cook my chicken for salads using a Chinese method. It is succulent and so tasty.

1 chicken

3 cups water

1 cup soya sauce

1/2 cup dry sherry or white wine

3 to 4 slices fresh ginger

1 shallot chopped

1 teaspoon salt

Combine all these ingredients, except the chicken, and bring to the boil. Add the chicken, cover and boil for 30 minutes. Turn off heat and allow chicken to cool in the liquid. Remove and chop into desired pieces. The stock can be used again and again.

Pineapple salad with Stilton mayonnaise

2 egg yolks

1 tablespoon whole grain mustard

2 tablespoons wine vinegar

1 1/2 cups vegetable oil

1 cup crumbled Stilton cheese

salt and pepper

1 large pineapple, peeled, cored and sliced into long slivers

375 grams (12 oz) cooked, shelled school prawns

Combine the egg yolks, mustard and vinegar in a blender. Add the oil, pouring in a thin stream to make a thick mayonnaise. Add the cheese but don't blend it too smoothly. Season to taste. Combine the pineapple and prawns and arrange in serving dish or dishes. Like the preceding recipe this looks good in a lettuce-lined bowl. Spoon the mayonnaise over the top and sprinkle with paprika or chopped green herbs. Serves six.

Pineapple goes well in practically any salad you can think of, but I think it is stretching things a bit far to try and combine it with a classic Niçoise or Caesar and I for one do not want to eat Pineapple Tabbouleh. Proust, who mentions food so often and with such resonance, writes of a Pineapple and Truffle Salad in Volume III of *Remembrance of Things Past*. It was served to an Ambassador who ate it 'with the inscrutable discretion of a diplomat' without revealing whether he enjoyed it or not. Some American salads which feature pineapple would test the skills of diplomacy. Some are so sweet I am not sure whether they are meant to be eaten for main course or as a pudding.

If you wish to make pineapple jelly from fresh fruit you must first cook it. Bromelin, the enzyme present in uncooked pineapple, will prevent the gelatine from setting. Here is a recipe for a pineapple jelly which should be eaten as a first course with the addition of the Stilton dressing from the previous recipe or simply a good mayonnaise.

Pineapple salad mould

3 cups pineapple juice

juice of 2 large oranges, strained

2 wineglasses dry white wine

1 good tablespoon white wine vinegar

3 level tablespoons gelatine

6 tablespoons cooked pineapple chopped into dice

red capsicum

rocket

Combine pineapple juice, orange juice, wine and vinegar in a large saucepan. Heat until almost boiling. Take three quarters of a cup of the hot mixture in a basin and sprinkle the gelatine on the top. Let it soften and then stir until it is completely dissolved. Add to the rest of the juice mixture and pour about a third into a wetted ring mould of about five cup capacity. Place in the refrigerator until just about set, then add the diced pineapple and pour on the remaining cool, but still liquid, juice mixture. Return to the refrigerator and leave several hours or overnight to set. Unmould onto serving dish and fill the centre with slivered red capsicum and rocket leaves. Serve the dressing separately. The quantities of gelatine ensure a firm jelly suitable for tropical climates or summer in Australia and beneficial to the nerves of those who are a little frightened of unmoulding. Serves six or more if you add more salad garnish.

The next recipe for pineapple fried rice is meant to be piled into the scooped out pineapple shells, which in a way echoes the wonderfully carved vegetables which are sometimes used to serve food in Thailand. This is an ephemeral craft where art is often induced to defy nature. For instance, the outer surface of a watermelon may be carved into a pineapple, and a pineapple by contrast can be turned into a palm tree.

I went to a class in Bangkok. I love the little carving tools — six little golden-coloured 'scalpels' on a ring, but I would rather admire somebody else's efforts from a distance than create myself. Perhaps it was because of the subjects we had to prepare. The cucumber fish was terrific fun but the anthurium made with a tomato and a chilli was just too vulgar for words! Anthuriums in reality are a parody on plastic I think.

A good compromise for a serving vessel would be a lovely footed dish in the blue and white pattern of Thai china that is probably the most commonly encountered both here and in Thailand. This is the overall pattern of interlocked shield-like shapes, in fact a representation of the little round segments of the skin, with a flower-like dot in the middle of each 'shield' — the pineapple pattern.

Khao paad sapparot-fried rice in pineapple

1 1/4 cups Jasmine rice

1 good sized pineapple

125 grams (4 oz) green prawns, shelled, deveined and chopped

125 grams (4 oz) fresh crabmeat, shredded

2 Chinese sausages, finely sliced

1 shallot, finely sliced

2 small red chillies, finely chopped

2 cloves garlic, crushed

2 tablespoons vegetable oil

2 tablespoons fish sauce

2 tablespoons soya sauce

freshly ground black pepper

crisp fried garlic — available ready to serve from Asian grocers

Boil rice in plenty of salted water until it is almost cooked, but not quite. Drain and leave to cool. Cut the pineapple in half lengthways and scoop out the flesh. You will only need to use half the quantity of fruit for this recipe so store the other half for another recipe. Cut the half into very small dice. Heat the oil in a wok and fry shallot, garlic and chillies for a minute or two. Add the prawns and continue to cook, stirring until they turn red in colour. Add the crabmeat and sausage and stir to blend. Add the rice, fish sauce, soya sauce and pepper. Stir well and cook over a medium heat until the rice is hot and fully cooked — about three minutes. Stir in the diced pineapple and cook another minute. Pile into the pineapple shells (or a serving dish) and sprinkle with crispy fried garlic. Serves 4 or 5.

Brunch with place-mats from Vietnam and Thai pineapple china. The menu includes pineapple salad mould with Stilton Mayonnaise, muffins, baby upside-down cakes and juice

There are two names for the pineapple in Thai — sapparot and makhnat. Sapparot means many eyes — all the little individual fruits which make up the pineapple. There is a folk tale from the Philippines about the many-eyed pineapple — a strange plant which a childless woman grows in her garden. It seems this idea of the many eyes is an Asian observation.

Fruit is a sign of prosperity and can be seen on family shrines and altars in Chinese homes and shops all over south-east Asia. Pineapples were observed in profusion by my son and daughter-in-law in celebrations in Phuket, the beautiful island off the coast of south-west Thailand and paper 'Chinese Lanterns' in the shape of pineapples are tied up for New Year in Bangkok and Singapore.

When I visited Singapore recently I was very keen to track down some of these paper lanterns. The first I saw were at Bugis Village. The old, rather seedy Bugis Street, frequented by most beautiful kaitais or transvestites, up to 20 years ago, has been rather sanitised and transvestites apparently just don't exist in Singapore any longer. However good food does, and my husband, my son and I tried some delicious Peranakan or Straits Chinese dishes. The waiter who was one of the fattest men I have ever seen — quite monumentally huge — kindly told us where we could try to buy lanterns the next day.

It was the most colourful and exciting pineapple sleuthing of all time! On the second floor of a block of rather old-fashioned shops was a collection of merchants specialising in devotional requisites for Chinese. Pellegrini Chinoise. The first thing I saw were packets of the paper 'money' that is given and burned at Chinese New Year and other festivals. It is always beautiful, in colours of gold and red and I had in other parts of the world seen pictures of sages, horsemen, flowers and urns — but never before pineapples.

Then I saw pineapple candles and then the most extraordinary variety of ornamental pineapples — golden plastic ones, large and small, glass and plaster, and glorious ones made of folded paper money. Some of them seemed to combine incongruous elements—dragons, little people, fish and...leeks? I found it hard to decide and they were certainly not cheap, but in the end I succumbed and bought a plaster version of a red 'brick' stove surmounted by a 'bamboo' steamer from which erupts pink and white steamed buns topped by a pineapple. It is something like Botticelli's 'Birth of Venus'. But the vision splendid doesn't end there. Looking like currants, or the cloves with which one might stud the Christmas ham, are little lights within the 'diamonds' of the pineapple skin. When the whole creation is plugged in, these little lights flash on and off in red, blue, yellow and green. It is utterly wonderful and utterly awful. I love it.

I also bought the 'paper' lanterns but was told proudly that they were not paper but plastic and therefore able to be washed. I have small, medium and large, and several of each, so washday could be very festive. Some are yellow and some are orange and I have also a sequinned and beaded version to hang from the rear-vision

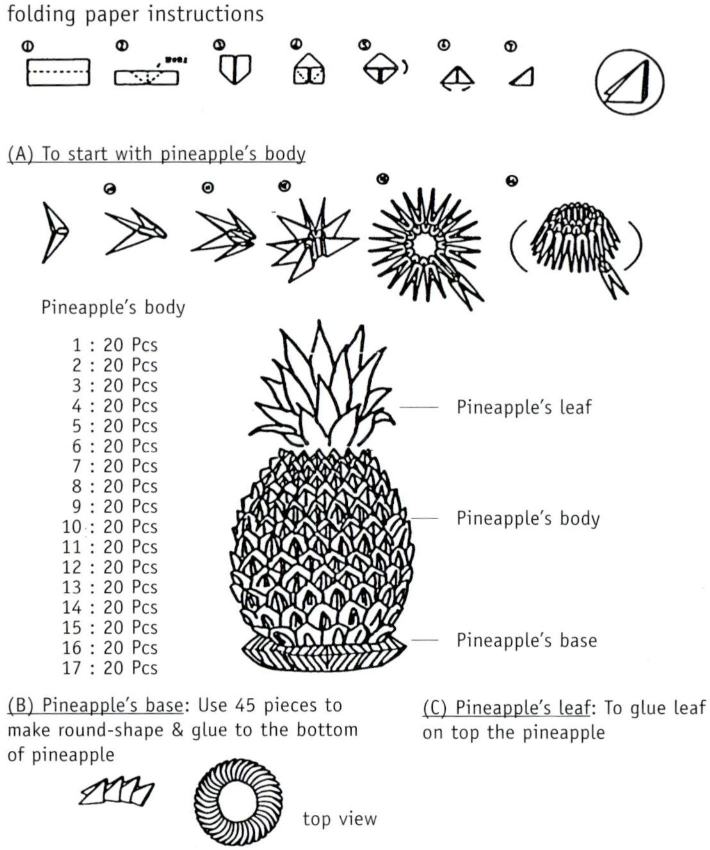

Origami pineapple

Paper lanterns

Pineapple candles from Singapore with eucomis or pineapple lilies

mirror in my car and little packets of origami paper to turn into a do-it-yourself pineapple.

The reason for all these glorious representations of the King of Fruit is that for the Chinese, particularly in south-east Asia, the pineapple has come to be the symbol of prosperity. The inscriptions on the paper money portraying the pineapple say 'money comes' and one explanation is that in Hokkien the word for pineapple, which sounds phonetically in English like 'ong lai', is very like the words that mean 'prosperity comes'. So much of Chinese symbolism refers to something which has a name that sounds just like the word for a virtue, or an emotion, or a condition. The monosyllabic nature of the Chinese language, in which the same sound can mean many different things, lends itself to a play on words. Decorative motifs are part of a symbolic, homophonic language that uses images to speak. It is a glorious system of visual puns and a logical fascination.

Probably the first exotic encounter most people of my vintage had with a pineapple was in the local Chinese restaurant, when a vividly pink platter of something in sweet and sour sauce was brought to the table as part of a family feast. Chicken and almonds, prawn cutlets, fried rice, beef in oyster sauce and sweet and sour fish were usually the dishes my family ordered at Chopsticks or Tow Yin. Fortunately the luminosity of that wild pink seems to have diminished in Australia these days but the reddest I have ever seen was in the Chinese restaurant in the old Bengal Club in Calcutta. Wow, it was enough to make anyone hyperactive for a week. Indian cooks are familiar with red food colouring which is used in tandoori preparations, but the cook in the Club was undoubtedly Chinese.

There are many Chinese in Calcutta and in the curious, seductive way of India, some embrace a kind of Hinduism. Many of the women work as hairdressers and the men as cooks. In the days before Bata, and *A Suitable Boy*, the Chinese shoe-shops all over the sub-continent were considered the best. There was still an active one in Peshawar when we lived there, but slowly they have disappeared. Good sweet and sour is very good and here is a very simple recipe.

Deep fried fish with sweet and sour sauce

500 grams (1 lb) firm white fish fillets
1/2 teaspoon salt
1/2 teaspoon finely grated green ginger
1 egg white, slightly beaten
vegetable oil
cornflour
Sauce:
3 green capsicums
1 medium onion
1 large carrot
1 bamboo shoot
2 slices fresh pineapple
3 tablespoons vegetable oil
3 tablespoons sugar
1 tablespoon soya sauce
1 tablespoon dry sherry
1 tablespoon vinegar
2 tablespoons tomato sauce
2 teaspoons cornflour
1/2 cup water

To cook the fish:

Remove skin and bone from fish and cut into short finger lengths. Sprinkle lightly with salt and rub in the fresh ginger. Mix the egg white into the fish and let it stand while making the sauce. Then heat about a cup of oil in a wok. Toss the pieces of fish in cornflour, a few at a time, until lightly but well-coated. Fry a few pieces at a time over a medium heat until just done. Be careful not to over-cook or it will become dry. Drain on absorbent paper. Arrange on a platter. Pour on the sauce and serve immediately with rice.

To cook the sauce:

Cut the capsicum and pineapple into 3 cm (1 inch) squares, the onion into lengthwise slivers and the carrot and bamboo shoot into thin slices. Heat the oil in a wok and stir fry the vegetables for a minute or two. Mix the sugar, soya sauce, sherry, vinegar and tomato sauce together and add to the wok. Stir in the pineapple. Blend the cornflour with the water and when the mixture in the wok is boiling, stir it in. Cook, stirring the mixture until it thickens and boils.

I read an article in the Qantas in-flight magazine on memorable tea shops around the world. One of the *salons du thé* The Tea Chapter in Singapore and it is very different from the others recommended. It was opened in 1989 by graduates of a Singapore management school who wanted to preserve the ancient tea ceremony as taught by the sage Lu Yiu.

Six kinds of Chinese tea are served—green, floral, lightly fermented, heavily fermented, black and white. The teapot used is of a rare purple clay and each person has two teacups, one for smelling and one for tasting. Three factors are considered when drinking tea — the colour, the aroma and the taste, and if all this is just too serious, one can also enjoy one of the specialities of the house — pineapple tarts.

My daughter, Emily, who flies to fascinating destinations as a member of flight crew for

Emirates Airlines, helped do some pineapple investigations. I had written to the Tea Chapter but hadn't heard back. So Em visited the tea house and explained her mission, asking if I might have their recipe. 'Oh, no,' said the assistant, 'Queen Elizabeth II has visited our teahouse and we don't give recipes, but you can buy a pineapple tart to take away.' She did this and also bought some from another shop. I tasted both, the latter were in the usual form for a jam tart — a little concave pastry shell with a glazed and sugared lattice of pastry strips over the top. The pastry was lard-based. Those from the Tea Chapter were little twisted bolster shapes with what tasted like pineapple jam in the middle of melt-in-the mouth pastry.

I experimented with them and came up with this recipe based on one for a Greek cookie. I used pineapple marmalade for the filling (see Page 29).

I was later delighted to receive a long fax from Kris Han, the Marketing Manager of the Tea Chapter, with the recipe for their open-face tarts and information about the history. Pineapples have been grown successfully for centuries on the Malay Peninsula, she wrote. There is even a town in southern Johore called Pekan Nanas or Pineapple Town. It was named for the pineapple cannery that used to be there. The factory has gone but the name continues.

Pineapple cookies

250 grams (8 oz) margarine
3/4 cup sugar
3/4 cup oil
grated peel of an orange
6 cups self-raising flour
3/4 cup orange juice
pineapple marmalade

Place a cup or so of marmalade in a sieve over a bowl. The filling for the cookies shouldn't be too moist. The strained liquid and any leftover jam can be put back in the jar or stirred into some ice-cream or made into a delicious sweet sauce.

Cream margarine, sugar and orange peel, then gradually add oil and continue beating until the mixture is light and fluffy. Fold in the sifted flour alternately with the orange juice. Knead lightly. Take dough by the scant tablespoon. Flatten into an oval in the palm of your hand and place a teaspoon of the jam in the centre. Fold the dough over to enclose it and shape into a neat oval. Stamp the top with the tines of a fork and place on greased oven trays in a moderate oven180°C (350°F) for about 20 minutes or until pale gold. Cool on wire racks.

This makes at least fifty rather hefty cookies. You can experiment and make them smaller and smaller to resemble the dainty variety as sampled in Singapore.

Pineapple tarts in many guises, including intricate, pineapple-shaped pastries with a clove stem, are now a traditional treat at the time of Chinese New Year and also at the Hindu festival of Hari Raya. Nobody seems to know why but 'someone at the National Museum' suggests that they have 'evolved from colonial rule when fruit tarts were a favourite snack of the ruling Brits'. Locals hit upon the pineapple as appropriate filling, the jam sometimes infused with cinnamon. Somewhere along the line they have become more associated with the Straits Chinese, or the Peranakans, with the lard-based, open-face versions being a popular staple in Malacca. Whatever the history, they are delicious.

Tea chapter tarts

600 grams (1 1/4 lb) plain flour

250 grams (8 oz) butter or margarine

2 eggs

1 teaspoon self-raising flour

1 regular size pineapple — grate & squeeze dry of excess juice

250 grams (8 oz) sugar

Stir dry, grated pineapple and sugar in a saucepan on a low heat and stir continuously to avoid burning. When just about dry, put aside to cool. Mix the butter and eggs. Sieve the flours together into a bowl. Make a well in the flour and pour in the butter and egg mixture. Mix well and knead. Rest the pastry in the 'fridge for half an hour then roll out and cut into shapes to fit into your tins. Rest again for half an hour. Place a small amount of the pineapple filling in the centre. You can add pastry strips as a lattice if you desire. Bake in a moderate oven 180°C (350°F) for 15 minutes or until golden.

I later visited the Tea Chapter myself — with my dear tea-drinking husband, and not disclosing that I am not mad about tea. What an enjoyable morning it was. The tea-house is situated in historical Tanjong Pagar, an area where pretty Chinese shop-houses have been conserved and restored most beautifully.

Because our visit was just a few days before the Autumn Mooncake Festival there were no Pineapple Tarts to be tasted but as snacks with our Imperial Golden Cassia tea — a light fermented Oolong and the type dispensed to the Queen in 1989 — we were served vegetarian pao or steamed buns and baby mooncakes. The reason for the vegetarian nature of the pao filling was that if animal fat was transferred from one's hands to the little porous clay tea-pot the delicate taste and aroma of the tea would be spoiled.

The teapots are tiny and so are the sweet little porcelain cups. The one used for sniffing is cylindrical in shape and the one for tasting is wide-mouthed and shallow and each contains just a few sips. The second brew is considered the finest and we had about six or seven. Other teas have such marvellous names as Golden Turtle, Aged Narcissus or Iron Goddess of Mercy. We sat with our legs recessed under a low table just a foot above the floor and enjoyed the calm and peaceful atmosphere.

Here is another recipe for a rather more European tart — a large one.

Pineapple & almond tart

Shortcrust pastry

90 grams (3 oz) margarine

155 grams (5 oz) plain flour — 1/4 should be wholemeal for an interesting texture

1/4 teaspoon baking powder

2 tablespoons milk

Filling

1 pineapple

100 ml (3 fl oz) brandy or rum

125 grams (4 oz) sugar

3 egg yolks

300 ml (9 fl oz) cream

200 grams (7 oz) chopped blanched almonds

To make the pastry rub the margarine into the combined dry ingredients, sifted if you like, until it looks like breadcrumbs. Add the milk and mix to a soft dough. Knead lightly, then put the dough into a plastic bag and leave in the refrigerator for half an hour. Take it out and roll on a floured board with a floured rolling pin to fit a 23 cm (9 inch) pie dish. Put the dish back into the refrigerator for another half hour. These rests in the 'fridge will stop the pastry from shrinking when it is cooked. Bake blind in a moderate oven 180°C (350°F) for about 10 minutes.

Meanwhile prepare the filling. Peel the pineapple and notch out the eyes. Cut the fruit in halves lengthwise and then cut the halves in half. Cut out the core then slice the four long sections into 2 cm (1 inch) slices. Soak in the brandy or rum for an hour then strain and reserve the juice. Arrange the drained fruit in a circular, overlapping pattern in the partially cooked pastry shell. Beat together the egg yolks, cream, sugar and the brandy or rum until well mixed. Pour over the pineapple slices and sprinkle the almonds on the top. Bake in a moderate oven 180°C (350°F) for about 30 minutes until the top is starting to brown and the custard is set. Remove from the oven and for ease of serving let stand for half an hour. It will be much easier to cut. It is very good cold. Serves 6 to 8.

With very little time to brood about it, I found this year that I had to go into hospital for a bit of a female rearrangement and removal of parts past their use-by date. I arrived for admission, feeling rather ill-at-ease, but was received by a bright and cheerful Malaysian nurse who looked at me and asked, 'Haven't you been in here before?' I certainly hadn't and then she asked, 'Are you a cook?' 'Yes,' I answered, 'but how did you know that?'

She had remembered me coming in to visit my dear friend, Vanessa Griffith, a year before. It just so happened that while Vanessa and I were talking, Gek, the nurse from Malaysia, came in to ask all those routine, very basic questions that one must answer after an abdominal operation. Poor Vanessa wasn't feeling too well, and when she was asked, 'Have you passed wind?' and the answer was a rather sad, 'No', I just broke up laughing.

I then returned with some samosa and sambal for her to eat, with the advice that these would probably help do the needful. A friend, whose beloved is Chinese, told me that quantities of clear soup full of chopped green ginger was very efficacious in this regard as well.

Gek, who is a wonderful nurse blessed with a phenomenal memory, asked me all about my cooking and I told her about the pineapple book. She said her mother, who is a marvellous Peranakan cook from Malacca, had a special fishball and pineapple dish, a little like laksa, and that her pineapple tarts were famous. She has recently visited Canberra and I have now tasted both delicious creations, plus some superb pork and chicken dumplings.

Recipes come at very unexpected times! This one is one of those for intuitive throwing in of this and that and is specifically without garlic. If you can't live without it you could add some to the spice paste mixture.

Gek's mother's fishball lemak

250 grams (8 oz) fish balls bought from a Chinese grocer
½ a good-sized pineapple, peeled, cored and cut into pieces similar in size or slightly smaller than the fish balls
2 tablespoons oil
2 onions, very finely chopped
1 tablespoon of finely sliced lemon grass
2 teaspoons galangal powder
2 teaspoons ground turmeric
10 candlenuts, smashed or finely grated
1 tablespoon of sambal ulek or other chilli sambal
salt (Gek's mother says Malaysian salt is different)
2 tins (800 ml or 3 cups) coconut milk
coriander for garnish

Place onions, lemon grass, galangal powder, turmeric and candlenuts in blender or food processor and blend to a paste. You may need to add a little water. Heat the oil and fry the paste with the sambal ulek, stirring until it is cooked and fragrant. Stir in the coconut milk and salt to taste and bring to the boil, stirring the while. Turn down the heat and simmer for five minutes or so. Put in the fish balls and pineapple pieces to heat through. Serve in soup bowls, garnished with coriander leaves. Serves 5 or 6.

All over south-east Asia fruits and flowers have been used as the models for beautiful objects made in ceramics, brass, silver and other precious metals, ivory and wood. In Malaysia, there is a series of postage stamps depicting fruit, including of course the pineapple. In my collection I have pineapples made of brass, clay, glass and wood. I also have two beautiful silver pineapples. The tiny one was given to me in Laos by great friend and pineapple supporter, Carmel Sullivan. She ran the Australian Clinic in Vientiane with compassion, skill, tact and also a riotous sense of humour. She now lives in Brisbane on the road to Mount Cotton, which was one of the early pineapple-growing areas in the north.

The occasion of her gift to me was a farewell party we gave for some Irish friends who were leaving Vientiane. Because the community was small, and most people were relieved of the worst of the household jobs by Lao helpers, parties often were wildly imaginative. We all tended to see each other very regularly and go to lots of dinners so we tried to make them a bit different. This night it was — yes, of course, it had to happen — a pineapple party.

Guests came in a variety of green and yellow clothes. One French couple looked rather staid until they ripped off part of their clothing to reveal matching shorts patterned with pineapples and positively shrieking 'Fruity'! I had blockprinted paper tablecloths with my Nepalese pineapple woodblock and there were pineapples everywhere. The most impressive was one I had bought at the Palaveck market. This was where my husband's anti-opium project was situated and the new little market-building had been opened by the King of the Hmong hill-tribe group a couple of weeks before. The pineapple was colossal. It was some sort of freakish manifestation with at least 30 crowns shooting up from several large fused fruits. It was quite fantastic and lasted for weeks, sitting beautifully in a great copper urn we had bought in Peshawar.

We ate pineapple in various forms and some courses were just made to look like the fruit. I seem to remember I made fish mousse in some bar moulds which have fluted ends. I poured the central cream-coloured (I hope it wasn't too yellow!) mixture into the centre and blocked off the ends for a little while before pouring the portion that I had dyed with green food colouring into the double-ended 'leaf' sections.

The only other time I have done a complete pineapple repast was when, out of the blue, a Canberra conference organiser asked me to do lunch and a demonstration for a visiting group. I couldn't imagine what I could cook better than anyone else, but then thought that no-one I knew would possibly do pineapples in toto. The group was made up of the spouses of urologists and they weren't put off by the strenuous pineappleness of their lunch, so that should be comforting. I guess I *could* have told them that the way to introduce oneself to the dubious joys of drinking one's own urine, according to self-imbibers, is to start off with pineapple juice. Is it the colour I wonder?

Well, Carmel gave me the dear little silver pineapple, and a year later in Saigon on Rue Catinat, immortalised by Graham Greene in *The Quiet American* and now known as Dong Khoi Street, I bought a bigger one. This is a most fascinating street to wander and I did so with Jamie and Emily. We had come by train from Hué where we saw a very unkind thing done to a pineapple. We had spent the day on motor-cycles visiting the Imperial tombs which lie quite a distance from the city. They were beautiful, but I must admit the day was tinged with terror for me, clinging onto the broad back of my son and trying not to gasp as I saw certain death approaching. It must have been just as bad, if not worse, for him, having a terrified mother impersonating a limpet on his back.

Happily we did return to the city unscathed and were utterly starving, a condition often induced by fear. So we went to one of the big Government-run hotels on the banks of the Perfumed River and, not wishing to spoil our appetites for the evening, we just ordered a quantity of nems, the delicious little spring rolls. The wait was quite lengthy and when they finally

High tea with Barking White Matriach Cake, Pineapple Fruit Cake, Pineapple Upside-down Cake, Pineapple and Almond Tart, Peranakan Tarts, Cookies, ice-cream and flummery

arrived we could understand why. The spring rolls came impaled, along with those little paper umbrellas which turn up now and then in one's life, into a poor pineapple. The crown of leaves was further decorated with gerberas and ferns, a kind of gilding of the lily. I have heard that in Hanoi, in addition to suffering the indignity of the impaling, sometimes the pineapple is hollowed out and a candle inserted in its base. A clear case for the Royal Society for the Prevention of Cruelty to Pineapples.

The train journey from Hué to Saigon was dreadful. To start with the train was twelve hours late but we couldn't walk too far from the station just in case it arrived. I did a lot of reading and ran out of anything edifying so had to try one of the children's new age sci-fi books in desperation. We hadn't been able to book sleepers so had 'soft' seats. We travelled a few hours into the night and I thought to myself that if indeed I managed to survive the journey I would be a physical and mental wreck. I really had to lie down. Fortunately at Da Nang three berths became available and that was a huge relief. Em and I shared a four-berth compartment with a honeymoon couple — I think they were exhausted, too.

In 1976 when the children were four and five we travelled all over India in second class railway carriages for two months and didn't think of discomfort. I thought it supremely comfortable in comparison with the third class carriages in which I had travelled when I was 21. What's more, in 1976, we just about always had booked seats so there was none of that awful gearing of one's nerves and muscles to hurl onto a train as it steamed into the station, hoping against hope that there would be a seat or at least a space in the luggage rack. The third class carriages no longer exist, and after my last unscheduled trip in a train from Bodh Gaya to Calcutta in 1985, I haven't wanted to catch any Indian train. It just shows you what happens with the onset of old age, and a slightly fatter wallet. However, for a real taste of India I think every visitor should have at least one short journey — and there are Imperial Maharaja-type carriages available on some routes now. All things are possible if that is your will and pleasure.

Eventually, a whole day after leaving Hué, we arrived at the Ho Chi Minh City Station where the cyclo drivers are much trickier than in Hanoi. We set off in two to find a decent hotel and from the guidebook selected a likely one in Dong Khoi Street. It was utterly revolting, rat-infested, smelly and dirty, so we dashed off again in two new cyclos to the comfortable Rex Hotel where we recovered.

We saw lots of sights and went to the Cuchi Tunnels and to a most marvellous congregational worship with fascinating music at the Cao Dai Temple at Tay Ninh. Rue Catinat, however, provided a surprising and worthwhile insight into the techniques of post-impressionism, an unexpected Saigon pleasure. Walking from the river one day we turned a corner and saw, in the window of a shop, an 'original' van Gogh. It was one of the studies of 'Iris' which, of course, are well-known to many Australians after the Alan Bond saga. We went in and there, in a corner, was a young woman painting an 'original' Salvador Dali, while other artists were executing other paintings we recognised. I asked the price of the 'Iris' and I can't quite remember what it was, but I was directed to go to the other studio up the street where there was another. They were so beautiful I thought that it might be just the acquisition I needed — a van Gogh from Saigon. The proprietor of this shop wasn't there so I had to return later in the afternoon. When I asked him again for the price he told me something much higher. I haggled long and hard to just get it back to what I had been told earlier. To cut a long story short, I went back the next day and the next and he wouldn't budge. There was a Gauguin of a similar size also for sale at a much lower price. Why, I asked was the Gauguin so much cheaper than the van Gogh. 'Oh', he answered, 'van Gogh's are much harder to paint than Gauguins'.

Years later in Singapore I was eating Penang Asam Laksa for the first time. I had eaten Laksa Lemak many times before — that superb dish of noodles in a coconut milk soup — but never this sharper, sourer, more interesting version. My son James was with me and I asked him to have a spoonful to analyse the flavour. We agreed on the tamarind and the pineapple, which was the reason I had chosen to eat it. Suddenly he said, 'oh, this is just like that Vietnamese soup, you know, *canh chua cá*.' I hadn't at that stage tried *canh chua cá* so I couldn't agree or disagree. I later discovered he was absolutely spot on — the two are very similar. I have experimented with an amalgamation and think my version is just as good, if not better than either! But there is a problem now of nomenclature — what do I call it?

Sour fish & noodle soup

500 grams (1 lb) firm fine-fleshed fish — fillets or cutlets
5 cups water
1 $^{1}/_{2}$ cups tamarind water
8 stalks Vietnamese mint
sugar, salt and pepper to taste
500 grams (1 lb) of laksa noodles — I use dried, coarse vermicelli
2 teaspoons blachan or trasi (shrimp paste)
6 dried chillies, soaked in hot water until soft
2 stalks lemon grass chopped finely
$^{1}/_{2}$ teaspoon turmeric powder
15 shallots, chopped
1 medium tomato cut into half and then into slim wedges
1 $^{1}/_{2}$ cups pineapple cut into 3 cm (1 inch) wedge shapes
1 Lebanese cucumber cut into 3 cm (1 inch) pieces
1 cup bean sprouts
1 small brown or purple onion sliced into rings
saw-leaf herb or coriander
crispy fried onion or garlic

Poach the fish in 3 cups of salted water for 10 minutes or until cooked. Remove the fish and flake when cool. Keep aside and retain the stock.

Wrap the shrimp paste in aluminium foil and fry in a dry frying pan for a minute or two. Remove the foil and put paste in a blender with the chillies and their soaking liquid, lemon grass, turmeric, shallots and one cup of water. Blend until pureed.

Pour this mixture into a saucepan and add the remaining 2 cups of water, the tamarind water (made from soaking tamarind pulp in hot water, squeezing it and forcing it through a strainer — tamarind stock cubes are also now available) and the Vietnamese mint. Boil gently until a good aroma is evident and then add the fish stock, sugar, salt, pepper, tomato and $^{1}/_{2}$ cup of pineapple pieces. Simmer gently for 30 minutes then remove the sprigs of Vietnamese mint and put in the flaked fish. Heat through. While this is all happening cook the noodles. Bring a large saucepan of salted water to the boil and then put in the noodles. Bring back to the boil and cook uncovered for one minute. Then remove from heat, put the lid on the pan and let it sit for about 8 or 10 minutes until the noodles are soft. Strain and place servings of noodles in individual, deep bowls. Arrange the remaining pineapple, cucumber, onion and bean sprouts on top of the noodles and pour on the soup. Garnish with chopped saw-leaf herb or coriander and crispy fried onion or garlic. Serves 6.

Saw-leaf herb is used in cooking all over south-east Asia and is sometimes referred to as long coriander. Its botanical name is *Eryngium foetidum*. It is a long, rather tough green leaf with a serrated edge and is a member of the thistle family. I first discovered it in Trinidad where it went under the curious name of Shadow Benny. This was a corruption of Chardon Benit, or Blessed Thistle, and was also called dhaniya by Trinidadian Indians — the Hindi for coriander. The flavour is just about exactly the same as real coriander (*Coriandrum sativum*) which did not grow easily in the West Indies.

I would never have known about it then but for the enigmatic advice of our eccentric Indian gardener, Sonny. He was about 5 feet (1½ metres) tall and about 70 years old and rode an ancient motorbike at about 24 kms (15 miles) an hour which caused one of our neighbours to call

Pineapples by the door. The painting at the top is of Shadow Benny or saw-leaf herb or long coriander or dhaniya or eryngium foetidum

him Hell's Angel. He was often drunk and had simple methods of horticulture. Once a marvellous pink frangipani had caterpillars so his cure was to cut it down. I rushed outside in horror when I saw what he was doing and said, 'Sonny, whatever are you doing?' 'De worms Ma'am,' he answered, 'de worms'.

Gardeners are often drinkers. The one we had in Vientiane was always in one of two conditions — drunk or hungover and the only reason we kept him on was his wonderful name, Mr Hoo Wat!

Shadow benny is available in markets all over south-east Asia, but in Port-of-Spain it grew as a weed. In the perpendicular garden it was found in hazardous cliffs and I would pluck it while clutching overhanging roots — yes, really! We later found a wonderful supply in a graveyard and made regular journeys there. I became so fond of it and all the funny connections to harvesting it in Trinidad, that I asked a well-known botanical artist to paint me a picture of it. She said it was the first weed she had ever been asked to immortalize.

The silver fruits that could be bought in the morning market in Vientiane were from Vietnam, but the charming little animals all came from Cambodia. Some of these little containers, for they all opened up as a box, must originally have been paraphernalia for the ritual of preparing and chewing of betel, but now they are being created just as ornaments.

The morning market or Talat Sao in Vientiane was open all day, as was the evening market. I did most of my shopping for fresh fruit and vegetables either at the big Talat Thong Thum, or when in a hurry, at the friendly little Talat Simuang. There was a great variety of everything, wonderful green herbs and flowers to buy by the bunch. The Lao eat a lot of flowers too and they would appear amongst the fruit and vegetables. I rarely bought beef. It was terribly tough and the vendors tended to cheat on weight. I suppose I could have tenderised the beef with pineapple or papaya but I didn't. The only beef I ever bought was put through the rather terrifying Russian mincer I had bought at the Talat Sao.

If I had bought steak to grill, a traditional Vietnamese accompaniment would have been Mam Nem.

Mam nem – fish & pineapple sauce

1 teaspoon sugar

1 tablespoon water

4 tablespoons fish sauce

2 tablespoons fresh pineapple, finely chopped and squeezed dry

1 fresh red chilli, finely chopped

Dissolve the sugar in the water in a small saucepan over a low heat. Allow to cool then add all the other ingredients. As well as being good with a grill, a spoonful is wonderful in congee or soup.

When we arrived in Vientiane there were many Russian goods in the market and many Russian people buying everything else. This was before the disintegration of the United Soviet Socialist Republic in 1991. Most of the Russians in Laos were not permitted to fraternise with any other foreigners and one could not even meet their eyes in the street. Gangs of women used to come into town from their compound in a battered old bus. They would descend on the, then, only existing little 'supermarket', sweeping past any other customers who happened to be in their path, to their ultimate goal of sausages and cheese in the freezer. They fell upon it like the wolf on the fold.

It must have been close to Paradise, or at least the Riviera, living in Vientiane at the time and not in Moscow. Cheap, attractive clothes in the market, no queues for good, cheap food and a warm climate — what more could one need? A new Embassy was being built and even blue-collar workers had been brought from Russia to construct it. They weren't all spies and scientists.

My son, James, was with us in Vientiane when the regime in Russia changed and he has in his collection of amazing things, the brass plate from the gate of the Embassy. Well, I guess they didn't want it. Hordes of Russians went home, but some interesting ones stayed. The bread scene changed when some of the women, freed from restrictions on what they could and could not do, started to make and sell delicious dark loaves. The Lao baguettes are splendid, but rather lacerating to the roof of the mouth, so an occasional Russian loaf was a welcome change.

Natalie Kerbo, one of my friends who stayed on, told me that soon after her arrival in Vientiane, she had to attend a lecture given to newcomers by the doctor at the Russian Embassy. He told them that at all costs one should avoid eating pineapples. The acid would destroy the enamel of the teeth and would also ruin the stomach. Little was said about malaria, dengue, parasites and the usual tropical health problems. Natalie said that this sort of advice was often given as anti-capitalist propaganda, as a safeguard against the desire to eat pineapple and other luxury items when back home in Russia after a tropical sojourn.

As long as there wasn't a scare about trichinosis the pork in Vientiane was good, but the most entertaining food to purchase were chickens. Some people I knew bought them live and then might ride home with them tied by their feet, dangling from the handlebars of their bicycles. I bought them despatched and plucked. The vendors always seemed to have well-manicured hands with long painted fingernails and golden rings. They would thrust these hands into the chook and pull out the giblets and the embryonic, as yet, shell-less eggs. I would shake my head at these and they would be put aside for other customers as they are considered a great delicacy. The best part was something that you probably weren't meant to see. If you looked behind the vendor, poking out of her basket was a bicycle pump. Perhaps you might think it was just to ensure the incident-free trip home. But no, this was an important part of the poultry seller's equipment. The pump was used to inflate the chicken so it would look wonderfully fat and plump and appealing to the would-be buyer.

I used to buy coconut meat already shredded from one market stall which would go into this delicious Indonesian curry. It is very quick to prepare so is useful when unexpected visitors arrive at mealtime.

Pacari

1 medium pineapple, not too ripe

1 tablespoon vegetable oil

1 small onion, finely chopped

1 clove garlic, crushed

small stick cinnamon

3 cloves

3 cardamom pods — bruised

2 teaspoons ground coriander

1 $^1/_2$ teaspoons ground cummin

$^1/_2$ teaspoon chilli powder or a slurp of chilli sauce

1 teaspoon salt

1 cup thick coconut milk

1 teaspoon jaggery or brown sugar

Peel and core the pineapple and cut into chunks. Heat oil in saucepan and fry onion, garlic and whole spices over medium heat until onion is soft. Stir to prevent burning. Add the coriander, cummin, chilli and salt and stir for a minute or two until spices are browned. Add the pineapple and stir to coat with spice mixture. Add the coconut milk and sugar and bring to the boil, stirring constantly. Simmer gently, uncovered for about three minutes or until the pineapple is just tender. Serve with rice, a meat or fish curry or what you will. Serves six as a side-dish.

We had many visitors in Vientiane, many of them interested in textiles, but one came in search of food and she was Rosemary Brissenden. We are all so used to books on Asian cookery now that we tend to forget that it was Rosemary who did all the pioneering work. After visiting Indonesia in her student years and finding the food irresistible, she researched and wrote the first edition of *South East Asian Food* in the 1960s which Elizabeth David, our great guru, suggested was 'a book which every serious cook should possess'.

When I met Rosemary in Canberra in 1992 she was about to launch into the mammoth task of revising and expanding her book, which had by this time become an out-of-print classic. She had never visited Laos so I invited her to stay. It was very interesting to take her about the town, looking at things from a different viewpoint. We visited Talat That Luang, the market on the edge of the town where most of the really exotic wildlife was sold, much to the discomfort of the World Wildlife Fund people. Little pathetic bundles of fur — rats, bats or squirrels were sold everywhere, including on the steps of the Post Office, but the wildest range of game was to be found at the market — snakes, termites sold in piles on banana leaves, birds of different species, many of them very beautiful, frogs, pangolin and deer. There wasn't a single animal which existed in Laos that people didn't eat.

Rosemary watched so carefully when we wandered through the markets and looked at the food-stalls. My friend, Michel Somsanouk, who could answer any question I might pose about art, craft, history, literature...and food, took us to breakfast at a famous noodle soup stall in a backstreet. Rosemary asked questions as to subtle differences in both taste and technique that I had scarcely noticed. It was a very interesting exercise for me to observe.

The new edition of her book is a great success and she very kindly agreed to talk to the book group of which I am a member in Canberra. Each month's gathering of the Bookworms is such a stimulating and enjoyable exercise, where members discuss whatever it is we have read. The group is small, only eight members, and the rule is that half are to be Australians and half from foreign countries. We are all different, which means we don't always agree and this generally leads to good discussion.

A feature, which was meant to be minor, but because it is such fun has become less so, is that, without any competitive baking, whatever we have to eat to conquer literary-induced hunger must relate to the book. We have had a superb Viennese chocolate treat called 'The Black Moor in the White Shirt' to go with *The Moor's Last Sigh*, home-made bread and honey with *Remembering Babylon*; Violet Crumble Bars and meat pies with tomato sauce with *The Songlines* and 'Aunt Daisy's Lemon Pudding' with *The Stone Diaries*.

We didn't critically discuss Rosemary's book when she attended. Instead she told us about its creation and cooked a wonderful lunch. She has allowed me to reproduce here one of her pineapple recipes. This is a red curry from central Thailand which traditionally appeared only at special occasions such as important temple ceremonies, weddings and ordinations.

Kaeng Khua Mu Sapparot-Pork & Pineapple Curry

1 small pineapple

250 grams (8 oz) pork shoulder, cut in bitesize pieces

2 cups thick coconut milk and 2 cups thin coconut milk

$^1/_2$ to 1 cup water or chicken stock

2 tablespoons fish sauce or to taste

Paste spices:

1 tablespoon coriander seed

$^1/_2$ tablespoon cummin seed

6 large dried chillies, seeds intact, soaked in warm water for 30 minutes and then squeezed dry

1 stalk lemon grass, finely sliced

2 teaspoons kaffir lime skin, chopped

3 tablespoons garlic, sliced

4 tablespoons shallots, sliced

2 teaspoons shrimp paste

To make the spice paste, roast the coriander and cummin seeds in a dry pan until they are brown and aromatic. Grind them to a powder in a mortar. Add the chillies, lemon grass, lime skin, shrimp paste, garlic and onions, pounding after each addition until you have a fine paste. Peel the pineapple and shred by slashing with a sharp knife down to the core, holding it and turning it in your other hand. Set the shredded fruit aside. Put one cup of thick and one cup of thin coconut milk into a saucepan and bring to the boil. Stir occasionally as it boils until it has evaporated considerably and the oil begins to float on top—about 10-15 minutes. Put in the curry paste and mix thoroughly. Turn the heat down and stir as the oil continues to come out until the spices smell cooked and aromatic. Add the meat and stir thoroughly. Then add a little of the water or stock. Stir well and simmer until the meat is half-cooked. Add the pineapple to the curry and then the rest of the thin coconut milk, and the water or stock. Simmer without covering until everything is well blended and a little reduced. Stir in the fish sauce and leave the curry simmering gently for about an hour, or until tender. Add the remaining thick coconut milk. Bring back to the boil and serve with boiled rice.

In Canberra I can't grow pineapples but my mother gave me a pineapple lily bulb from her garden in Sydney, and despite the fact that it looks beautifully tropical, it weathers the fierce winter and flowers through the summer. Its botanical name is *Eucomis* and it has greenish flowers in a dense raceme on a leafless scape crowned with a tuft of leaves. The stem and leaves at the base are smudgily spotted and it lasts weeks if picked and put in a vase in the house.

I also have rapacious pineapple mint but I find it utterly useless from a culinary point of view, and its position in the garden is because of its pretty, soft, green colour and its ability to fill up space.

My plantation of pineapples in Vientiane was in a bed beside the clothesline. I used to think how pretty it all looked when I had a line full of hill-tribe textiles. I used to buy numbers of Hmong belts and baby's hats. The baby's hats are so

Eucomis or pineapple lilies. The Greek name means beautiful hair. This implies a good head and refers to the tufted green crown of leaves above the flower spike which looks rather like a pineapple. They are native to South Africa

pretty, like the ultimate in party hat, designed so that lurking evil spirits will mistake the baby for a flower and pass by. They were always extremely dirty and I would wash them very carefully, minding the fugitive dyes didn't transfer themselves to undyed parts. I probably have the world's largest collection of belts — what a claim to fame! They are so beautiful and so are the hats, but I used to unpick those and use them in some of the jackets that I created for sale.

At first I found that my jacket styles were being copied in Vientiane so I looked further afield than the superb Lao handloom silk. I went to Cambodia and bought wonderful ikats there and usually I would combine a Lao textile on one side, reversing to Cambodian. Some featured a Hmong baby's hat as pockets or Indian sari border as cuffs, a piece of Afghan tashumar embroidery as a yolk or an intact piece of an old, ruined pha biang or Lao shoulder cloth as a complete back. I sold about 200 or so, in Laos, in Australia and in America, and it has been fun to see them on unknown people now and then.

One textile which would not have been quite suitable for my jackets is the beautiful piña cloth from the Philippines. This is the thin, light, transparent and often magnificently embroidered cloth made from the leaves of the pineapple.

The colourful textiles of the ethnic minorities in the Philippines have usually been of more interest to hunters and collectors of cloth than the fine textiles of the lowland Christians, but the story of piña cloth is as interesting as the history of the pineapple itself.

Another foreigner in the service of Spain, the Portuguese, Ferdinand Magellan, arrived in the islands we now know as the Philippines in 1521. He claimed the archipelago for Spain but was killed only weeks later in a battle against the natives on Mactan Island. Antonio Pigafetta, a nobleman from Venice who accompanied Magellan, wrote an account of the voyage. In descriptions of the natives encountered in the Visayas and Mindanao, he mentions tattoos, golden ornaments, cotton cloths embroidered with silk, silk cloths and other 'cloth woven from the bark of a tree'. It is clear that the natives of these islands knew how to weave long before the arrival of the Spaniards.

Colonial occupation began in Cebu in 1565 but shortly after this the settlement moved north to the island of Panay in the Visayas group. Here the natives were friendlier and the soil more fertile. Piña was first produced here and today it is still the centre for production. One would presume that the Spanish either brought planting materials or else crowns from fruit eaten on board ship took root accidentally after being thrown away.

Other more complicated but intriguing possibilities for the origins of the first plantings exist. There had long been important trade links with China when junks early travelled the South China Sea, the Sulu Sea and the coasts of Borneo and Mindanao. They were the forerunners of the famous galleon trade which flourished between 1565 and 1815. The galleons, so redolent of swashbuckling adventure, focused on the ports of Canton, Manila which was established as the Spanish stronghold in 1571, and Acapulco in Mexico. The Spanish administered the Philippines from Mexico until 1821 and goods were brought overland from Vera Cruz in the east on the Gulf of Mexico to Acapulco in the west on the Pacific coast. Perhaps pineapples came that way. I certainly didn't know this when I drank pineapple juice at sunrise at Vera Cruz on Holy Saturday in 1985.

Yet another theory is that the Chinese brought pineapples from Peru in 1592. They were certainly being grown in China in 1594. But the very fact that the fruit is called piña and not any variation of the word *ananas* indicates the arrival took place at least during the Spanish period, and most probably because of Spanish enterprise.

Manila became an entrepot port where Asian luxury goods and fine fabrics were exchanged for Mexican silver. Competition with Chinese textiles caused some traditional Philippino cloth industries to slump but others developed in imitation of imported goods. The Spanish encouraged weaving and particularly embroidery. The styles employed had their origins not only in Europe, but in India, Arabia and China where embroidery is said to have been 'invented'. Ordinances forbade natives from wearing imported cloth and there were regulations regarding dress. These rules had an important effect on local handlooms.

Religious schools and convents called *beaterios* were established by Spanish missionary nuns to teach not only religion and the other three 'r's, but also sewing and embroidery. The pupils were girls from wealthy families and orphans alike. They were taught to ornament garments, altar cloths and other church vestments with the finest stitching which rivalled that of anywhere else in the world. This fine work was successfully exported through the galleon trade and later when the port of Manila was opened in 1834, to international traders. Thus the textile industry survived and improved and in 1848 at least 52 different varieties of fabrics were being produced.

Although once woven by the Torajan people of south Sulawesi and perhaps by the Chinese, fine piña cloth has always been associated with the Philippines and particularly with the island of Panay. It is hardly surprising that people who had discovered they could weave cloth from such unlikely things as banana or abaca, would experiment with the strange new pineapple. This they did, after the arduous extraction process, on simple bamboo looms. At first the town of Iloilo on the south-east coast was the centre of the industry, but today it has moved to the northern part of the island, in an area around the town of Kalibo.

Nicholas Loney, the vice-consul for Panay, wrote from Iloilo in the 1860s that there were 60,000 looms in use, producing piña, jusi or silk and abaca cloth. Loney actually contributed to the subsequent decline of the industry as outrider of the industrial revolution and introducer of sugar as a huge, new cash crop to the Visayas. Cheap factory-made textiles from Manchester were making inroads into the local market and Loney promoted sugar as a return cargo. The 60,000 looms produced 39,673 piezas of cloth for export in 1863. In 1869 the total had dwindled to 12 700 piezas and by 1873 only 5,100 piezas were exported.

Iloilo today is a bustling city and busy port but the clack of the bamboo handlooms is scarcely heard any longer. Along the coast at intervals of about 10 kilometres there are immense Spanish churches. At Miagao, which used to be a piña-weaving town, the church has a strange and beautiful pediment. It shows, in relief, St Christopher and the Christ Child amid luxuriant tropical vegetation. There are laden papaya tress, the very essence of fecundity, equally fruitful stylised guavas, and in the centre like a palampore tree of life, a tall and sheltering coconut palm. St Christopher is holding fast to it as if it is a shield from one of the typhoons that visit the island regularly. The church, built between 1787 and 1797 holds a congregation of about 1,500 and the architecture is protective and massive. It looks almost like a fortress and it has actually fulfilled that role. One could really sing with conviction, 'A mighty fortress is our God', within its walls.

The church-dotted coast reminded me of the imposed European presence in Ghana — punctuating castle-forts instead of fortress-churches. There was even an allusion to slavery in the seventeenth century watch-towers by the sea at Guimbal. These were used as lookouts for Moro pirates who came to capture natives to sell as slaves in Mindanao.

Despite the Declaration of Independence in 1898, when the Philippines, free from Spain, became the first republic in Asia, America paid US$20 million to the Spanish for their former possession and only granted full independence in 1946.

The American era brought a revolution in fashion and a further decline in handloom fabrics. Embroidery however flourished and lace-making became another important and lucrative occupation. Needlework, as in Spanish times, was taught in schools to avoid sweatshop situations and to give an impetus both to the arts and for girls to supplement family incomes. American demand for embroidery increased and cloth, such as pearline lawn, was imported. It was easier to handle than piña and was available in commercial quantities.

The 1920s lifestyle changes rang a death-knell for native costume, which had been much influenced by Spanish dress. The Charleston and bicycle-riding required more freedom than Philippino clothing allowed. By 1938 most Philippinas were wearing American vestidas which were not particularly easily made from locally-produced handloom cloth.

World War II caused another setback, but the 1950s saw a resurgence of a demand for embroidery. By the mid-'70s the weaving of fine piña had dwindled to a dangerous level. Former First Lady Imelda Marcos urged mechanisation of industry, but with little result, and as all the production went to her, there was no raising of awareness among the buying public. Moreover, many people did not wish to be associated with her efforts.

In 1986 the Patrones of Casa Manila, a group of lovers and promoters of traditional Filipino culture, expressed concern at the situation of piña production and Patis Tesoro, an extraordinary woman of boundless interests and energy, made a number of field visits to Kalibo to size up the situation. She found only about 300 metres of cloth was produced per month by a diminishing number of weavers, all of whom were elderly.

There are several fine collections of wonderful old piña in the Philippines. I was fortunate enough to examine pieces belonging to the Bangko Sentral ng Pilipinas—neckerchiefs or *pañuelos*, barongs or men's shirts, blouses and baby clothes. The embroidery was magnificent, with the finest stitching applied to the gauzey cloth—*sombrado* or shadow-work, *calado* or drawn-thread and beautiful flowers, foliage and scrolls applied in many different stitches, some so fine it was hard to see. As well as pieces made from pure piña cloth, some garments were of piña seda or piña and silk, some with a silk warp and piña weft, some with suksok—inlaid stripes or scattered motifs of supplementary weft known as pili.

I was also able to view the Bank's wonderful art collection. Philippinos are incredibly artistic and paintings even figured in the struggle for freedom. When Juan Luna and Felix R. Hidalgo won First and Second prizes at the Madrid Salon of 1884 their talent and rewards were seen as an assertion of independence and national pride.

The earliest European influenced art is found in religious paintings. The vestments of the saints and clergy are probably piña copies of Spanish lace and this is partly what piña garments did in general. They reproduced locally, with indigenous flair, the models from Europe.

More interesting than the saints in the quest for piña is the collection of portraits of illustrados. These are nineteenth century pictures of the respectable bourgeoisie—illustrado literally meaning the enlightened. Dressed in their most beautiful clothes and jewellery, they are icons of status and comfortable pride.

Most of the sitters, male and female, are dressed in clothing made from piña. The finest artists of the time reproduced in miniaturismo style the delicacy of the embroidery on the filmy cloth.

Men wear barongs and women mostly the Maria Clara style of dress popular in the late nineteenth century. This was named for the ill-fated heroine of the book *Noli me Tangere* by the national hero, Jose Rizal. She was a model of respectability and propriety and the illustrado portraits mirror her virtues.

Leticia Jimenez, whose portrait I am permitted to reproduce, sat for the painting by an unknown artist from Santa Ana, Manila in the late nineteenth century. Leticia looks demure with eyes downcast. Her hands are crossed and one finger touches a book, probably a missal, indicating her piety. Her *baro* or blouse has wide, open sleeves, the shape of an inverted tulip and is surmounted by the *pañuelo* or neckerchief which is edged with a scalloped design and fastened with an *alfiler* or pin.

Leticia also wears ear-rings, a *paeneta* or ornamental comb and two *agojillas* or hairpins. It was usual for women to wear a *tamburin* or chain around the neck from which would dangle a pendant or locket containing a miniature picture of a saint, the Virgin Mary, Agnus Dei, a sweetheart if betrothed or one's father if not. We cannot see Leticia's chain and she wears but one unostentatious ring — some portraits show women with up to eight.

There are several family portraits and one is of Cirilo and Severina Quiazon and their children painted by Simon Flores. It is a large and wonderful picture which incorporates a snatch of landscape from beyond the open window, a still-life in the glass verina on a table, details of an

Portrait of Leticia Jimenez, painted by an unknown artist from Santa. Ana, Manila in the late nineteenth century. From the collection of Bangko Sentral ng Pilipinas

illustrado house interior, as well as four charming portraits. Their gazes are level and they look comfortable in their beautiful clothes.

One change has been made to the original work and that is to the chubby baby in his mother's lap. Alluding perhaps to his parents' down-to-earth attitude to life, he was painted without pants. The baby grew to manhood and learned how to smoke cigars. He also grew tired of all the teasing about his infant penis and one day he burned it out with his lighted cigar. The painting was restored with baby fully clothed.

Piña handkerchiefs and fans were often shown in the hands of the portrait subjects. Neither were just beautiful ornaments but had a language of their own and were coyly used by ladies in the art of flirtation and as a means of unspoken communication. More than anything else piña handkerchiefs had international appeal. They were a finishing touch to a well-bred lady's wardrobe, a vehicle for artistic expression on the part of the embroiderer and a perfect gift, even for Royalty. One was given in 1862 as a wedding present to Princess Alexandra. They were keepsakes carried by men going off to war or on long voyages — with a sweetheart's name perhaps embroidered in one corner. Sometimes human hair was the 'thread' used for the stitching, echoing mourning jewellery from Victorian England which incorporates locks of hair.

Philippinas look wonderful in baros and pañuelos and modern updates of traditional dress. In the Centennial Year of Independence, Government servants are urged to wear Filipiniana on Mondays and there is a resurgence in wearing national dress. Large and lumpy foreigners do not necessarily look so good in such clothing so the importance of the handkerchief lingers on. These days it is one of the most affordable pieces of piña one can buy.

Men too carried them in the illustrado portraits. Perhaps they didn't have pockets in their clothes just as the barong tagalog has none today. This traditional men's shirt has never undergone the decline in popularity that so affected women's national costume. In three weddings I watched in the Church of San Agustin and the Manila Cathedral one Saturday afternoon, there was only one man not wearing a barong in all the extensive gatherings of families and friends.

The reason I was at San Agustin in Intramuros, the oldest part of Manila, was to see the superb pulpit of gilded wood. It is poised above the congregation under the high vault of the ceiling painted with illusionistic architectural detail and is thoroughly beautiful. The crowning touch is a wonderful finial in the shape of an upside-down pineapple. Though, I ask myself, can a touch be crowning if it is at the bottom?

Golden pineapple finial on the pulpit in the Church of San Augustin, Intramuros, Manila, Philippines

It was inspiring to meet many Philippinos who were intensely patriotic and proud of their nation and culture. In particular I met three marvellous women—Frances Arespacochaga at the Central Bank, India dela Cruz Legaspi who lives at Kalibo and Patis Tesoro. It was by luck that I met them and one passed me on to the next. Their kindness and interest was overwhelming.

Patis Tesoro is a Manila couturier, but that is just one part of her busy life. She designed the exposition of Philippino culture at the Musée de l'homme in Paris last year, she masterminded the production of the Ramos Presidential inauguration where 5000 sat down to eat together and, with former First Lady, Amelita Ramos, she co-founded the Katutubong Filipino Foundation. This is a non-profit organisation devoted to the revival and promotion of indigenous Filipino culture, arts and crafts in ways that are environmentally sound and economically productive. It is also designed to bring awareness and understanding of the country's rich heritage to today's generation.

We visited Patis in the ordered chaos of her home-cum-office-cum-studio-cum shop in San Juan. Stepping in from the green, leafy garden was like entering a beautiful bee-hive. Seventy people were cutting, stitching, embroidering and fitting in the workrooms; pretty little white dogs were here and there; glamorous gowns hung in racks; antiques and paintings lined the walls. There were capiz-shell shutters and finely carved *calado* or traceries which gave glimpses to places beyond and *aparadors* or cupboards bulged with magnificent piña and other textiles. Patis, the Queen Bee, somehow knew where everything was.

Katutubong Filipiña Foundation's activities include a joint venture with Coca Cola to save the knowledge and usage of natural dyes from extinction. The project is centred in the Province of Abra in northern Luzon. Here indigenous communities have benefited immensely, not only through the creation of jobs, but also through the establishment of green forests in former barren wastelands. They are full of sources of natural dyes. As I found in Bangladesh, the lovely green pineapple leaf dye is also not know in the Philippines so I have sent that recipe off to Katutubong. Achuete, turmeric and coconut husk are just three of many dyes which have been refined by the foundation.

When Patis visited Kalibo on her mission to save piña, things looked distinctly gloomy but it all has a happy ending. The Fibre Industry Development Authority and the Aklan State College of Agriculture in 1990 set up a training centre to assist the Industry. Farmers are given agricultural advice and newcomers to the industry are trained as extractors and weavers.

There are now 300 weavers instead of the 30 of twelve years ago and approximately 8000 people are involved in the production from planting to weaving. Ten thousand metres of piña fabric are produced a month in a variety of forms. As well as the pure piña there is a piña bastos which is a coarser version; the mixture of piña and silk called piña seda; poly-piña which is a synthetic mix and abaca pinukpok or fine abaca.

I met Dr Helmar Aguilar, the President of Aklan State College of Agriculture, who was instrumental in starting the revival. The College at Banga, about 7 kilometres from Kalibo is surrounded by lush, green forest. When the piña project started, Dr Aguilar was faced with the problem of not having enough planting materials. It was a real dilemma but he hit on the perfect solution. In the Philippines, beauty contests, where the winner is crowned Miss this or that, are very popular. Dr Aguilar organised a Miss Piña Contest where the winner was not necessarily a great beauty, but the provider of the greatest number of planting materials. Families and friends scouted up mountains and into forests to find crowns, slips and suckers which were growing wild. More than 100 000 were collected to start the project in fine style.

Most piña plantations are small-holdings of about 20 or 30 metres square. The plots are not weeded in the later stages as the tall grasses and weeds help hold up the long pineapple leaves which grow to between 100 and 200 centimetres (1-2 yards). The variety suitable for textiles is the Spanish Red or Native Philippine Red.

The leaves take about 18 months to reach maturity and a sign of their readiness is that the fruit on the plant is also ripe. The leaves are snapped from the plant with a sharp tug on the tip, the thorny edges are pulled off and then the fruit is plucked as well. The fruit is smaller than the other main variety grown on Panay, the Smooth Cayenne or Hawaii, as it is known locally. It is sweet and juicy and is generally eaten raw or made into jam. These days, though, with the piña revival in full swing, there is precious little time for jam-making.

Leaves once plucked can be left for up to three days for extraction. When this takes place, extractors squat on the ground by a long board on which the leaf is placed. It is scraped with a broken china plate — antique porcelain is best though, understandably, becoming a little scarce. A few inches on the leaf base is folded over and struck with the heel of the hand and the first set of fibres is revealed. These are the bastos or coarse fibres which are pulled off and set aside. The leaf is scraped again, this time with coconut shell, and the finer linawan fibres are extracted.

Extracting piña bastos fibre, Panay, Philippines

Classification for most textile fibres takes place later but the pineapple leaf is extraordinary in having these two distinct sets of grades. Bastos fibre account for 75 per cent of that extracted and linawan 25 per cent. Bastos formerly was used for the cascading, curly hair of images of saints and the Holy Family in churches and shrines around the country. Once I knew this I was quite compelled to see as many as I could — some of plaster, some beautiful old ones of wood with ivory faces which seemed so lifelike. Bastos fibre was also used for making twine and paper but since the refinement of piña bastos weaving there is little raw material left over. Some waste still goes to paper-making but most is fed to the family pigs.

When a quantity of leaves has been stripped, usually 1000, the fibre is washed in a stream and scraped again, this time gently with a sea shell, to remove all traces of green matter. The now white strands are partially sun-dried on the grass, then wrung and gently beaten for a few minutes with a bamboo stick to separate the fibres. It is then hung on a line to dry, combed with an ordinary hair-comb and tied upside-down to a slim bamboo pole ready to be knotted. Good eyesight is required here to tie the fine fibres together. The ends are cut off with a sharp piece of bamboo and the thread coiled in a clay pot containing sand to prevent tangles. Farmer-extractors take their fibre to the traders in Kalibo where the weighing system dates from the Spanish era. The weights are old silver coins.

India dela Cruz Legaspi runs her textile business from an old wooden building in Kalibo which

Washing piña fibre, Panay, Philippines

used to be her father's legal chambers. On the outside, as well as the flag of the Philippines, there was a banner to Our Lady of Guadeloupe. It seemed like a nice link with the place where Christopher Columbus first encountered our friend the pineapple.

As well as spear-heading the revival of piña weaving, India is an artist, specialising in delicate paintings on fine piña cloth. She had just returned from a visit to India, her nameplace, as a delegate to an Asian cultural symposium. She, and her husband Javier, were very kind and we beetled around Kalibo in their private yellow tricycle, a motor-bike with side-car. Perhaps it is India's artist's eye that guides her experiments with new dyes, designs and mixes. Piña is going to new overseas markets and innovative work is seen at home even in the church. We saw a wonderful red chasuble in an exhibiton of ecclesiastical vestments in Manila. It was made from cloth which incorporated piña paper threads in the weave.

Across the road from India's office was a building housing a number of her looms where piña seda was being woven painstakingly. Elsewhere we saw pure piña inching slowly forward. The maximum production for a weaver in a workshop is half a metre a day. At home, where domestic duties intrude, daily production is about half that. All weavers must be patient but piña weavers must be almost saintly. It is so slow and they must constantly knot broken threads. Holy pictures are fixed to many looms, so maybe prayer is part of the art.

Looms are no longer made of bamboo. It is not strong enough and movement would result in uneven weave. The overhead frame loom is made of coco-lumber and the two foot-treadles are made of bamboo. Thread is warped on

Weaving piña cloth — Kalibo, Panay, Philippines

Warping piña thread prior to weaving — Kalibo, Philippines

cylindrical frames and it takes about a week to set up a loom. Thread for the weft is wound on bobbins made of small pieces of bamboo.

After weaving, the off-white coloured cloth, which has a rather smoky smell, is washed in rice water or citrus juice. It used then to be calendered or beaten to produce a sheen, but this is no longer common. Cloth to be embroidered is sent north to Luzon where women in Batangas and Laguna work their artistic magic. Lumban is a centre where breathtakingly fine work is created.

If one person performed the whole laborious process of extracting the fibre and treating it, then setting up the loom and actually weaving the cloth, it would take four months' solid work to produce 20 metres of cloth. Attempts have been made to mechanise piña weaving but the thread is so fine and prone to breaking that it is not feasible. It is extraordinary that the finished cloth, as fine as gossamer, is so strong and durable. Not surprisingly, piña is expensive and beyond the reach of most Philippinos.

Rites of passage, requiring special clothing and regalia, are important to other ethnolinguistic groups over the country. They are important to lowland Christians and piña has assumed a degree of ritual signigificance to those who can afford it. The revival has ensured that vestments, christening robes, wedding gowns and dresses for such events as Presidential inaugurations, will continue to be made from the beautiful piña cloth which had seemed so dangerously close to disappearing.

The pineapple occupies a prominent position in imagery in the Philippines. It is seen in colonial furniture and architecture, in ceramics, wood-carving and fine, old silverware, especially in ornamental tooth-pick holders called *paliteras*. But, with the title of my book in my head, I was most delighted and astonished to find there are no fewer than 46 riddles, some with five or six variations, where 'the answer is a pineapple'. Several focus on the multiple 'eyes' of the skin of the fruit, but many more use the idea of a house surrounded by swords, thorns, nails or even guns. The house can be that of a king, queen, princess, beautiful lady, priest or a whole legion of saints — St Mary, St Anne, St Helena, San Miguel. 'I wear a crown, but I am not a queen. I have scales, but I am not a fish. What am I? The answer is a pineapple'.

The Torajans from South Sulawesi wove pineapple fibre for ritual clothing and screens to partition sleeping areas for guests. Cloth for the Torajans has a status like that of holy relics. I was present at a funeral in a village near Rantepao. Buffalo had been slaughtered *en masse* and great numbers of people had arrived to join in the ceremonies which involved quite some catering. Cloths had been wrapped around the posts of verandahs and draped the coffin. The daughter of the deceased person received us. She wore a beautiful sarong and Robyn Maxwell told me later, when she looked at my photographs, that it was from Flores, really quite a long way off.

I used to work as Robyn's volunteer at the Australian National Gallery when she was curator of Asian Textiles. Her *Textiles of Southeast Asia* which was published in 1990 is a work of great scholarship. It is one of my most treasured books and one of the most frequently consulted. Many of the Gallery jobs I did were pretty mundane but nevertheless enjoyable and I became quite an expert at making the conservation-safe reels on which precious textiles would be rolled. Sometimes such a mechanical process frees the mind to wander elsewhere. We had lots of laughs and I occasionally performed other more responsible tasks and handled some magnificent pieces. I learnt a great deal.

I have two pineapple Christmas tree decorations which were bought in India. They are golden orbs with a wonderful cockade of green tinselled leaves and look a little like a multi-fingered hand — a kind of King Midas hand in reverse. The way they hang from the tree is upside-down which is a very Indian touch and presupposes the pineapple upside-down pudding.

This time-honoured pineapple pudding or cake I feel sure came to the Antipodes from America. Though, as people from the northern hemisphere are so fond of telling us — and heaven knows why they think it so funny — we are the ones who are upside-down and therefore perhaps we invented it, just like the Pavlova. And, who was it — the New Zealanders or the Australians?

Shonagh Koea from New Zealand was another writer at the 1997 Canberra Word Festival and was a tremendous discovery for me. I had noticed her on the first day as she has such a strong and interesting face. I went to one of her sessions the next day and was entranced. At the conclusion of her talk I rushed off to buy the book of short stories from which she had just read. They were all sold but to my delight there was a novel partially set in India, *Sing to me Dreamer* which I have now read and enjoyed immensely. I did want to buy the short stories as there was mention of an upside-down pudding served to two widows

Decorations on a Christmas tree and the pineapple in the garden as ornament

in the middle of a tempest when the entire universe seemed upside-down.

When the author autographed my book we had a bit of a chat about India. She had spent some time with her husband staying in the Palace of the Maharajah of Jaipur and this was fuel for the novel. I did ask her about the pudding and she said that her sister, who was ten years older, always cooked things that to her as a child used to the usual stuff we all ate in the 1950s, seemed the ultimate in sophistication. One of these things was pineapple upside-down cake with the rings of pineapple, tinned of course, and the glacé cherry in the centre of each golden ring, and she remembered how it was always a drama to turn out.

Jane Grigson, in her fruit book, rather uppishly tells one to 'resist the temptation to stick glacé cherries into the holes of the pineapple slices'. She suggests lightly toasted hazelnuts rubbed free of their skins would be more harmonious. I have to say to dear Jane, God rest her soul, that that is bilgewater (no, not Jane Gardam author of *Bilgewater* who was actually speaking at the same session as Shonagh Koea) and that the classic pineapple upside-down cake absolutely *must* have glacé cherries. If you are in mourning perhaps a substitute could be a stoned prune, but that touch of red in the cherry must be the perfect garnish.

I learned how to make this dish when I attended Sydney County Council cookery lectures at Crows Nest when I was eighteen, but I seem to have lost the recipe. I think I had decided, that despite its utter 1950s resonances of sophistication, I would never make it again. I did make it then for my younger brother and sister. I wonder if they have treasured the experience.

This recipe is an amalgamation of several. The addition of Angostura Bitters is definitely a Trinidadian touch. In the Caribbean, and particularly Trinidad where the Bitters is made, a splash is added to practically everything — meat dishes, soups, cakes — everything. It is also a universal cure-all for anything that might go wrong, especially stomach-aches. The recipe is a closely guarded secret but is a blend of herbs and spices and probably had its origins in Venezuela. We once spent Easter in Venezuela and ate the most delicious thing several times. They were called *golfiados con queso* and were hot, sticky cinnamon buns with cold, salty cheese rather like fetta. Oh yum...but here is:

Pineapple upside-down cake

210 grams (6-7 oz) margarine

1 1/4 cups sugar

4 eggs, separated

1 1/2 cups cornflour

1 1/2 cups plain flour

1 tablespoon baking powder

1 cup milk

2 splashes Angostura Bitters

pinch salt

1 small pineapple, peeled, sliced into 1.5 cm (1/2 inch) thickness and cored

glacé cherries

Melt 60 grams (2 oz) margarine with 1/4 cup sugar and pour into a 25 cm (10 inch) cake tin. Brush the mixture up the sides of the tin. It will slide back down to the base and this is what it is meant to do. Arrange the pineapple slices, you may need to have semi-circles as well as circles, with a cherry in the middle of each. Then, have all your ingredients at the ready. I like to beat the egg whites first as it saves having to wash the beaters twice, but you mustn't let them sit around waiting. So, beat egg whites with a pinch of salt until they are stiff, but not dry. In a separate bowl beat margarine, sugar and egg yolks until creamy. Sift the dry ingredients together and add to the batter alternately with the milk and bitters. Carefully fold beaten egg whites into the batter and pour over the fruit in the tin. Bake in a medium oven for about 50 minutes or until golden. Test with a straw. Remove from the oven, loosen around the edges with a knife and turn out carefully. The fruit might stick to the bottom but is easy to rearrange. The fruited surface, the bottom, which is now the top, should be a lovely golden brown. If for some reason the pineapple looks a touch anaemic, place under a hot grill for a minute or two. Watch for burning.

You can also use this recipe to make individual upside-downs. Use metal moulds and cook for only about 20 minutes. I have to admit they are rather classier than the big one, but all taste delicious. Serve either size as a pudding or a cake with cream or yoghurt.

Another classic of the same era as Pineapple Upside-Down Cake is Pineapple Fritters.

Pineapple fritters

1 small pineapple, peeled and sliced in 1 cm ($^1/_2$ inch) slices — halve the slices if desired and drain in a colander

1 cup plain flour

1 dessertspoon baking powder

pinch salt

1 egg, separated

$^1/_2$ cup milk

vegetable oil

sugar

Sift dry ingredients together. Beat the egg white until it is stiff and set aside. Beat the egg yolk until slightly frothy, add the milk and stir into dry ingredients. Carefully and lightly fold in the egg white. Have the oil piping hot in a frying pan. Dip the *dry* slices of pineapple in the batter mixture and fry in the oil until golden. Dust with sugar and serve immediately with vanilla ice-cream.

The best pineapple dessert, I still think, is pineapple ice-cream.

Pineapple ice cream

1 large pineapple

300 ml (9 fl oz) pineapple juice (see page 25)

250 grams (8 oz) sugar (or less to taste)

rind and juice of a lemon or lime

300 ml (9 fl oz) whipped cream

a good and generous slug of rum

Peel pineapple and put in the blender with a little of the pineapple juice. Blend but do not purée too smoothly — there should be a nice texture. Place with the rest of the juice, sugar and the finely grated lemon rind in a saucepan. Bring to the boil and simmer for 5 minutes. Remove from heat. Add the lemon juice and freeze for four hours or until mushy. Whisk the mixture with a fork, then fold in the cream and rum. Pour into desired moulds and return to freezer until set. Serves 6-8.

I have a pineapple-shaped mould for which I sent coupons and money to, I think, a gelatine company, about 25 years ago. I just loved it then but I don't use it now. I certainly have quite a collection of 'pineappleabilia' now given me by kind friends, and in particular my children. Jamie and his wife, Amanda, have given me most beautiful little Thai ananas teapots and Em, in her amazing onslaughts on the world's shops, has sent me cake from Hong Kong, table napkins from England, Thai ornaments, Mexican napkin rings, and other magnificent trophies. Another collector's item is a splendid bath sponge in the shape of the pineapple, given me in Nairobi by Val Buss. I love it, but I hide it when we have visitors in case they think I have become obsessive. Of course I haven't!

I did dream of them last week. Pineapples I mean. And they were confused with religion, so what would Freud make of that? I can't remember whether I was there or not or whether someone was telling me the story in order to impress me with his erudition, but St Paul, after his encounter with God on the road to Damascus, asked for a piece of pineapple. I woke up saying no, it can't be, it can't be — they didn't know about pineapples in the Holy Land then. What a pedant I have become. I think it does reflect the fact that I've just read *Cockles of the Heart* by Marion Halligan for the Bookworms, and heard a good lecture on St James of Compostela, and I do worship at the parish church of St Paul at Manuka.

The fresh pineapples I bought in Laos, apart from the Palaveck giant, were small and fitted prettily on various trays and platters and bowls to act as flowers before consumption. The farmers at Palaveck one year grew a fine crop of pineapples, but sent them off to market without a thought as to how they would travel. The fruit was hurled willy-nilly into the back of a big six tonne Isuzu truck and by the time the 190 km (120 mile) journey was over the consignment of fruit looked a little like pineapple crush. Well, I suppose one learns from one's mistakes.

Remember those wonderful sweet-smelling, cool Pineapple Crush Bars that sprang up in Sydney in the 1950s? When we were taken into town in the school holidays, usually to go to the dentist, but often ending up with a terrific lunch at Cahill's, those sugary oases, seemed like the near-ultimate in desirability, especially in summer. Less sophisticated, but just as desirable, were the pineapple paddle-pops we had as a special treat in the car, a hunch-back Holden, coming back from the beach in the summer. Sticky, salty, sandy little bodies and those delicious ice blocks—Mum always had a chocolate heart which cost 7d. Paddle-pops were only 4d.

My little brother was once sick into his hat coming back from the beach and didn't tell anyone. What an amazing little stoic. His name is Maitland which was the oddest in the trio, but growing up in the 50s with names like Claudia, Maitland and Meredith, when everyone else was Susan and Robert and Jennifer was a bit of a trial. Now, of course, it's different.

When I was talking to my brother about the question of pineapples, he said, 'Well, don't you remember Bill Peach?' I may well write a book about peaches one day but the pineapple is the fruit in hand at the moment. 'No, no,' he said. 'Don't you remember Maitland Pineapple?' In a final three or four minutes of an edition of 'This Day Tonight' on television in the late 1960s, Bill Peach interviewed an Afro-American entertainer who sang a number for the television audience and his name was really Maitland Pineapple! My brother, Maitland, cannot remember the title of the song. He was heavily into the Beatles and Ravi Shankar at the time.

My son, James, has a musical connection too. He has sung with a group in Bangkok called 'Diplomatic Incident' but he tells me that there is another rather more famous one in the United States called 'The Pineapples from the Dawn of Time'.

James sent me a cutting from a Thai paper about a sensational fruit salad. I guess it was in connection with my pineapple quest but at the time he sent it I was researching a lunchtime lecture I gave at the National Gallery on Lord Ganesha and other elephants — 'Trunk-calls to the Gods' — and the article neatly covered both. The Elephant Lovers' Foundation threw a party for the elephants of Kanchanaburi, not far from Bangkok, and their dessert was a combination of 2400 stems of sugarcane, 2700 bunches of bananas, 100 watermelons and 1200 pineapples.

I don't need to tell you how to make a fruit-salad of normal quantities, but a delicious one I made in my earliest years as a housewife was a combination of crushed pineapple, tinned mandarin segments, coconut, chopped marshmallows and cream. It was called 'Ambrosia'.

You could serve any pineapple dessert with a small glass of Pineapple Wine. In the Caribbean, wine and other potent brews are made in many households. I used to make sorrel which is a ruby red drink made from a member of the hibiscus family, that which we call rosella in Australia. I made it as a soft drink but when fermented it could pack a nice little punch. Sweet potatoes, grapefruit and all the citrus fruits were used and then there was ginger beer and amazing drinks based on the local rum. Mona Irvine's alchoholic sorrel and Ralph Hoyte's ginger beer linger in my memory.

I didn't try the pineapple brew until I was living in Nairobi, more than ten years after leaving Port-of-Spain, but the recipe came from the *Express* newspaper in the latter city. I think the sub-editor for the day had been down to the test kitchen to sample some of the brews because the heading for this recipe was 'PINEAPAPLE WINE!' The wine *is* very sweet but my best brew wasn't too bad at all. The best part about it was watching the raisins shooting about on the second day like jet-propelled water beetles. I see now that I mustn't have chopped them or else how could they have shot around the bowl legless? I didn't have an earthen-ware bowl but I advise for entertainment value, a perspex salad bowl, tightly covered with cling-wrap.

Pineapple wine

2 medium pineapples

$1\,^{1}/_{2}$ kg (3 lbs) sugar

juice of 1 lime

$2\,^{1}/_{2}$ litres (2 quarts) water

15 grams ($^{1}/_{2}$ oz) yeast

500 grams (1 lb) raisins

Peel pineapple, then grate, squeeze and strain. Dissolve sugar in water and add to the pine juice. Add chopped raisins and yeast. Pour into an earthenware jar, cover tightly and let stand for 21 days to ferment. Strain and bottle. Improves with age.

A West Indian legend from the earliest indigenous inhabitants, the Caribs, says that at the beginning of time, man was wandering, starving, through the steaming savannah. The god, Kabo Tano, took pity on him and created a great tree from which were hung all the fruits and foods he could need — cassava, coconuts, sapodillas, bananas, *pomme cythere*, guava, mangoes — everything. Every man cut himself a branch to plant in his garden and so it is today that food grows near every Caribbean home. In our garden we had cassava, avocado, shadow benny, our thief-attracting pineapples, and mangoes that obligingly fell into our plot from our next-door neighbours' tree. But all that beneficence did not add up to a balanced diet.

The market in Port-of-Spain really wasn't my favourite place but we used to go there most Saturday mornings. The car that we bought when we first arrived was the worst I have ever had. I found it absolutely terrifying, but the day that the gear stick came right out of the floor fortunately happened to be when my husband was driving down Charlotte Street to the market, so I suppose I should be grateful. He thought it was funny — I didn't.

In the market the vendors were on platforms raised above the purchasers and this literally gave them the upper hand. There was no bargaining here. The pineapples generally came from Guyana and they were described, not as fresh, or juicy, or even ripe, but pretty. 'From Guyana darlin'. 'Pretty, pretty pines.' Being called 'darlin' in Trinidad, I should point out, does not constitute any degree of familiarity.

Indians had command of the lettuce and watercress section where water seemed to run continuously. Corbeaux, which were rather spectral members of the vulture family, waited for scraps outside, but the person I remember most clearly was the man who sold rat bait. He wore opaque dark glasses and seemed to have rather rodent-like teeth himself. His packets of bait were attached to a pole he carried as he walked slowly about, calling out 'Rart Beht. Rart Beht.' We were half-inclined to buy some but we were put off somehow by his ratty persona.

We did have a terrible 'rart' problem in the roof of our house in Cascade. They galloped up and down nocturnally until I went to the chemist and bought some poison. There were various kinds, and there was a brand I just couldn't bring myself to buy. It was called 'The Last Supper' and on the paper envelope there was a parody of Leonardo's painting with Our Lord and His Disciples displaying horrible little rat faces! The Trinidadians have a wonderful sense of humour. The next year at Carnival, Peter Minshall's band had the Rat Race as their theme and as usual he had designed the most stunning costumes.

We arrived in Trinidad in October and people said, 'Well, you must be looking forward to Carnival'. That's what they would have said if we had arrived in any month from June to February. Perhaps for March, April and May they may have said, 'Oh what a pity you've just missed Carnival'. We politely answered yes, we were looking forward to it, but were rather mystified by the whole thing. The language was incomprehensible. What was *jouvet*, and *Dimanche Gras* and *las lap* and *playing mas*? Of course we would learn, and what an extraordinary explosion of ephemeral creative activity it was. There was construction of incredibly elaborate and complicated costumes in band headquarters; the Calypso tents where the latest, often highly satirical, popular songs were aired; and the panyards where musicians played the steel drums. The very air was charged with the creativity of these funny, artistic and clever people. We made the best local friends there of any of the countries in which we have lived.

West Indian Chicken

1 chicken, jointed

juice and grated peel of 1 lime

salt and pepper

oil or ghee

2 medium onions, chopped

2 medium tomatoes, peeled and chopped

2 tablespoons raisins

2 tablespoons chopped olives

1 teaspoon hot sauce or ground chillis or a chopped, fresh chilli

1 clove garlic, crushed

1 small pineapple, peeled, cored and chopped into chunks

1 cup pineapple juice (see page 25)

Marinate the chicken pieces in the grated peel and juice of the lime, salt and pepper for an hour or so. Remove from the marinade. Heat oil or ghee and fry the chicken until golden. Then add the onion, tomato, raisins, olives, chilli and garlic. Saute gently until onion is soft. Add the pineapple juice. Stir and cover. Simmer gently for about 20 minutes or until the chicken is tender, turning occasionally. Remove the chicken pieces and put in the pineapple chunks. Stir to combine and cook for a few minutes. Check for seasoning. Pour over the chicken and serve with boiled rice. Serves 5 or 6.

We lived in Trinidad, but my husband's territory for his job with the Food and Agriculture Organization extended to Guyana and Suriname on the northern coast of South America. Guyana was formerly British Guiana and Suriname was Dutch Guiana and had been swapped by the British with the Dutch for Manhattan in colonial times, when European powers played with other countries as though they were chess pieces. Some bargain! Paramaribo, the capital of Suriname, had a strange atmosphere. The old town is a tense little Dutch settlement of colonial houses clutching each other as they would in Delft or Rotterdam or Amsterdam, facing north towards the sea and turning their backs on the Amazonian jungle and the whole continent behind them.

All of the West Indies is cosmopolitan, with many races making up each nation. Trinidad was almost half and half Indian and African with Chinese and white and other small minorities. The original indigenous inhabitants had been killed or removed by the Spanish before the arrival of the British. In Suriname it was even more exotic. Indigenous Amerindians of course still existed. Some of the original African slaves had escaped into the impenetrable bush in the seventeenth century and lived there undisturbed and uninfluenced for generations. They remain still a distinct culture from the other Afro-Surinamese.

As well as this distinct group, there were the Javanese who had come as indentured labourers, like the east Indians, to work for the Dutch colonial power. They had also kept a discrete culture which included their food. One could buy packets of ready-mixed saté sauce ingredients in the market, Javanese vegetables, saté sticks and *sambal olek*.

In neighbouring Guyana the economy at this time was in difficulty. The modern supermarkets were almost bare except for bananas, wooden clothes-pegs, brightly coloured bottles of sugary cordial and cassareep, a liquid made from cassava juice used in traditional stews and pepperpots. Hucksters in the market place sold all sorts of other foodstuffs that they had brought in their big, colour-striped plastic bags from other Caribbean countries. Bread in the hotels of Georgetown was made from rice-flour and was pretty awful, but by contrast there was very good Chinese food available from restaurants where some cooks from mainland China worked legally before making their way to the United States as illegal immigrants.

The dinner table's centrepiece is a plaster pineapple from Singapore, complete with flashing lights. The meal includes Pacari, Galloping Horses, West Indian Chicken, Chicken Salad, Khao Pad Sapparot, Ghanaian Roast Chicken and Chatnis

Pepperpots are the legendary stews of the West Indies. They were a kind of widow's cruse or never-ending food supply — a casserole that was never finished from day to day but was topped up with new ingredients. The tradition had started in the great houses of plantations and some were claimed to have lasted, unextinguished, for over 100 years. There is a story about a particularly good version eaten a day or two after the disappearance of a pet kitten, whose furry little body surfaced in the savoury casserole in due course.

I love cats so such an addition will not be found in my recipe on the next page.

Pineapple chicken with pillaw

1 medium chicken, jointed
4 tablespoons lime juice
4 tablespoons oil
2 tablespoons raisins
2 tablespoons rum
dash hot or chilli sauce
2 big tomatoes, skinned and chopped
$^{1}/_{4}$ teaspoon ground black pepper
1 tablespoon butter
$^{1}/_{2}$ pineapple, peeled, cored and cubed
soft white breadcrumbs

Pillaw:

2 cups rice
2 tablespoons butter
salt and pepper
stick cinnamon
2 cloves
2 bay leaves
2 cardamoms, bruised
water
dhaniya (coriander)

Rub chicken pieces with salt and marinate in lime juice for an hour or two. Heat the oil and fry the chicken quickly until browned. Reduce heat and cook further 10 minutes, turning occasionally. Add raisins, rum, chilli sauce and tomatoes, salt and pepper and stir to combine. Cover and cook a further 15 minutes over a low heat. To cook the pillaw, melt the two tablespoons of butter in a large saucepan. Pour in the rice, add the salt and pepper, cinnamon, cloves, bay leaves and cardamoms and cook, stirring over a low heat until the grains of rice become opaque. Then pour on five cups boiling water. Stir well and cover the saucepan. Turn the heat to low and cook about 15 minutes or until the liquid is absorbed. Turn off the heat. Meanwhile, melt the one tablespoon of butter, roll the pineapple cubes in breadcrumbs and fry to a delicate brown. To serve, arrange the pillaw on a platter and pile the chicken on to that. Pour over the sauce and top with the fried cubes of pineapple. Garnish with chopped dhaniya (coriander).

While we lived in Trinidad we visited Mexico which, if I hadn't already been in love with India, would have become No. 1 — such passionate history, vivid art forms and colour. And the best pineapple juice I have ever tasted was discovered at a little wayside stall just outside the city of Vera Cruz as the sun rose one morning.

We had arrived there at about 8:00pm on Good Friday to find there was no accommodation. My husband and my sister made me do all the negotiating, and with a week's worth of tourist Spanish, it wasn't so easy. After doing the rounds of dozens of hotels I was getting desperate until on a street corner a charming young man asked if he could help — in French. The result was that we followed him on his motor-bike, with us in our hired Volkswagen, right across the town to what was absolutely, definitely a brothel. It even had a red light. It did us for the night — but we left early the next day! Just outside the city there were dozens of little stalls sprouting a profusion of hanging pineapples and great big glass bottles of juice on the makeshift counters. We stopped to start refreshed.

Once, as a guide at the National Gallery, I had to take a group of American diplomatic wives on a tour concentrating on Australian Aboriginal art.

It was a pleasure as the women were charming and for me, learning about Aboriginal art had been one of the great prizes in doing the Guides course. The Ambassador's wife sent me a cookbook to say thank you. The book was one published by her family company and was on Mexican cookery. She has kindly allowed me to reproduce a recipe for Baked Pineapple from it.

Baked pineapple, natillas

1 large or 2 medium pineapples
sugar
3 tablespoons rum
1/4 cup margarine
Natillas custard:
2 cups light cream
1/4 teaspoon salt
1/4 cup sugar
1 egg and 2 egg yolks
2 teaspoons cornflour
1 teaspoon vanilla

Lay the pineapple on its side and cut out a thick slice, making sure you don't dislodge the leaves. Retain the skin of the slice. Carefully scoop out the flesh and cut into small cubes. Mix the cubes with sugar to taste, and stir in the rum. Put the cubes back into the pineapple shell. Dot with margarine and cover the whole fruit, leaves and all, with tin foil. Bake in a moderate oven of 180°C (375°F) for 20 minutes. Replace the sliced-off top and place on a platter. Serve with the chilled Natillas custard which is made by stirring together all the ingredients in a saucepan and cooking, stirring, over a low heat until it is smooth and slightly thickened.

Georgie Cleary, an adventurous young friend, was recently in Mexico City. She lived just near the huge statue that looms over so many of the scenes in Baz Luhrman's 'Romeo & Juliet'. She sent me a beautiful card of a folk dance from Oaxaca where women in brightly-coloured huipils or blouses dance in a row, holding a pineapple lovingly on the right shoulder. They must surely be a non-prickly variety. She also sent me some wonderful glass swizzle sticks with pineapples on the end and two little ceramic pineapple punch-bowl pots made at Santa Fe de la Laguna and San Jose de Gracia. I have consulted a book on Mexican folk ceramics and find that to make these often huge pots, the moulded leaves are placed side by side, covering the entire surface of the moulded form, in a process called sprigging.

We bought highly utilitarian, but superb, pots in a market in the town of Santiago Tuxtla on Easter Saturday. The town was small and clung to the side of a tall hill. It perhaps would have been unremarkable had there not been a huge Olmec head, dating from 1000 BC, resting in an enclosure nearby. Men were selling iguanas and there were huge mounds of bright red chillis. I think I can feel a trip to Mexico coming on.

I made tonnes of jam and chatni in Port-of-Spain and I realised why later. We had had to leave Jamie and Emily behind at school in Australia for the first time. It was so hard and I felt that they were so far away they might as well have been on the moon. It took them 55 hours to get to us in the school holidays. What a ghastly journey and at that stage neither of them liked flying much. How times have changed, for Em at least. After we put them back on the plane at the end of each holiday it would only be about half a day before I would find myself just having to make some kind of preserve. I didn't realise it for ages, but it was then, and still is now, the way I keep sane! And maybe it has been the common thread of the pineapple which has made the sanity more interesting.

Tin candlesticks from Mexico

A pineapple punch-bowl from Mexico

By nature of a postscript

Since I finished this little book and signed the contract, I have lived in Vietnam for a short time. Not in Ho Chi Minh City or Hanoi, but in Tra Vinh, a town of about 70 000 people, 203 km from Ho Chi Minh City on the Mekong Delta. When we arrived, we were not just the only foreigners in the town, but also the first to have lived there since the French left in 1954. This must therefore account for the fascination that we aroused as we, or rather mostly I, rambled around the town. People didn't just stare and point and giggle. They called out 'hoo, hoo, hoo' to attract my attention. Did they want me to stand there and gibber or dance?

I felt as though I was living in a third world zoo where visitors bang the bars and hiss and throw things to make the tiger roar or the monkey stand up and scratch its armpits. There is no concept of private space in a place where there are so many poor people living so close to each other, and the noise was almost dementing.

Life there for the non-working trailing spouse was extremely difficult, to put it mildly. The search for a house was more tedious than usual — we even had to go to the lengths of paying a builder to construct a bathroom and a kitchen as additions to the house we rented, as there were neither. But, Pollyanna, it had to be said that the pineapples there were extremely good.

The vendors sat under tatty umbrellas in the centre of the bustling market, a pale blue rubber glove on one hand cutting off the skin and notching in and out as our nineteenth century friend told us, like lightning. Most people bought them ready peeled with some of the stalk attached, like a very large paddle-pop.

Elsewhere in the market, I discovered a vendor selling the most beautiful little gelatinous moulded treats from a basin of cold water. Some of them had four colours, all perfectly delineated in moulds only about 1 cm (2 inches) long. When I first saw them I rushed home and returned with my pineapple-shaped moulds which she made for me beautifully the next day. Mr Quoc, who worked with my husband and spoke wonderful English, came with me so I could find out how she made them.

Vendors selling pineapples in the Tra Vinh Market on the Mekong Delta of Vietnam. Most purchasers buy the fruit ready peeled

Rau cau from the Tra Vinh Market in southern Vietnam. These jelly desserts are made from agar agar, sugar and coconut milk and coloured with natural ingredients

Her name is Ly Kim Phung and the jellies she made are called rau cau. The ingredients are agar agar powder, sugar and coconut milk. Agar agar jelly has the advantage of setting very quickly and also at a temperature of 27°C (80°F) which is much more practical to use in hot countries where many people do not have refrigerators.

I thought the colouring would have been artificial, but not so at all. Reds come from the water in which beetroot has been boiled and yellow and orange from carrots. I had noticed that the carrots here are so incredibly full of carotene that even washing them before scrubbing made the water orange and my hands after cutting them up looked nicotined. White is from coconut milk and browns come from coffee. Vanilla is sometimes used as a flavouring.

The green colour comes from the pandean leaf or screwpine which is pounded to a pulp and then diluted with water. It is a long, thin, spear-shaped leaf and its distinctive nutty flavour is used all over south-east Asia in desserts and sweet soups. In Vietnamese its name is *la dua*, or just *dua*, which oddly enough is one of the other names for pineapple. As well as *thom*, our versatile fruit is known here as *khom* and *dua*.

Phung, who was very shy, made a large pineapple mould which we ate on Christmas Eve when Jamie and Amanda came to stay. I made pineapple ice-cream to accompany it which was *raaaather* sweet. There is no cream in Tra Vinh, or unsalted butter with which to make it, or even evaporated milk. So I had to use the sweetened condensed milk that is one of Vietnam's staples. One can become addicted to *cafe sua*, a short hit of the strongest black coffee sitting on top of a great slurp of condensed milk in the bottom of the glass.

Ly Kim Phung, the rau cau maker, Tra Vinh, Vietnam

A pineapple rau cau on a blue and white Vietnamese plate. The yellow colour comes from carrots and the green colour from pandan leaf

Mona and Tom and four origami pineapples

As you have seen, I had certainly included Vietnam in the pineapple adventure. It was odd then to go and live there and consolidate some of the facts — and to do such funny things as visit Mr Hoang Hao's Gallery on Dong Khoi Street, Saigon to commission 'Mona & Tom', to be used as the cover illustration. I am so glad I have them, Mona and Tom that is, rather than a bunch of old iris. What's more, I have been able to make a beneficent bestowal to the Guides Room at the Australian National Gallery, of a 'Rembrandt' no less.

No-one would believe me if I say that the first word I learned in Vietnamese, and accidentally at that, was *thom*, pronounced *tom*, which means pineapple! But it is true. On the first day in Saigon, I found my way to the Ben Thanh market to have a bit of a reconnoitre in the crockery section. The blue and white bowls and plates are just wonderful, and very hard to resist. Then I saw some rather kitsch little fridge magnets which were labelled as menu holders. I thought the title rather amusing and I wondered if Mother should advertise at breakfast the bill of fare for the evening just so people could tick the attendance list, or perhaps never ever come home again. The magnets were in the shape of all varieties of tropical fruits and, of course, I was drawn to the pineapple, which I was told was called *thom*! So armed with my one word, when I found the section selling cutters and moulds I was able to buy the ones that my friend in the Tra Vinh market used so perfectly.

Fridge advertising

I saw for myself in Hanoi a poor pineapple hollowed out and holding a lighted candle inside, while speared with spring rolls on the outside. In Tra Vinh we ate *canh chua cá*, the sour fish soup flavoured with tamarind and pineapple, sitting in the street on little, low stools and spitting the fish-bones and tamarind seeds into the gutter like everyone else.

I read more books in two months in Tra Vinh than in almost a year anywhere else. I reread *The Vice-Consul* and *The Lover* by Marguerite Duras. To travel to Saigon from Tra Vinh one must cross the Mekong on the vehicular ferry at Vinh Long where *The Lover* begins. These books were much more disturbing on the second reading – perhaps because I felt I was going as mad as the mad woman from Battambang.

I also wrote more letters than in a year anywhere else – huge outpourings to dear friends, some of whom wrote wonderful replies. I wrote to Arundhati Roy and David Malouf, both of whom I met in New Delhi last year. I felt I had discovered an incredible link between *The God of Small Things* and a short story called *A Traveller's Tale* – twins and moths and dreams. I thought *The God of Small Things* was so wonderful. The first time I went to India was in 1968/69, the time in which the book is set and I went to tea estates. I went to Munnar, from whence the sky-blue Plymouth came, and Cochin, and what about making jam and pickle and canning...pineapple? And David Malouf even mentions the Big Pineapple in this same short story, *A Traveller's Tale*. I asked Arundhati Roy at the 1998 Adelaide Writer's week if she had read it. She said no.

Do you think there could be a course of study, or should I write another book, a sort of a sequel to *And the Answer is a Pineapple?* I will call it *Dealing with Depression on the Delta – the Pineapple to the Rescue*.

Selling pineapples near Ben Luc on the road to Saigon

Bibliography

Adler, Peter and Barnard, Nicholas
Asafo! African Flags of the Fante, London, 1992

Benson, Albert H.
Pineapple Culture, Queensland Agricultural Journal, Brisbane, 1903

Bidwell, Shelford
Swords for Hire, European Mercenaries in Eighteenth Century India, London, 1971

Brissenden, Rosemary
South East Asian Food, Melbourne, 1996

Burton, David
The Raj at Table, London, 1993

Claypole, William and Robottom, John
Caribbean Story, Book One: Foundations, Trinidad, 1980

Crowther, Geoff et al
Africa on a Shoestring, Lonely Planet, Melbourne, 1992

CSR, Australia's Home Made Jams & Preserves Book

Davis, Marian L.
'Piña Fabric of the Philippines', Arts of Asia, September-October, 1991

Dede, Alice
Ghanaian Favourite Dishes, Accra

Finlay, Hugh & Crowther, Geoff
Kenya, a Travel Survival Kit, Lonely Planet, Melbourne, 1994

Gillow, John and Barnard, Nicholas
Traditional Indian Textiles, London, 1993

Grigson, Jane
Jane Grigson's Fruit Book, London, 1982

Jacquat, Christiane
Plants from the Markets of Thailand, Bangkok, 1990

Kenya Museum Society
Museum Mixtures, Nairobi, 1989

Kingsbury, Noel
'The pineapple: King of British Fruit', *Journal of the Royal Horticultural Society* /the Garden Vol 120 Part 5, London, 1995

Lane, *Sunset Mexican Cook Book*, Menlo Park, 1985

Lotschert, Wilhelm and Gerhard Beese
Collins Guide to Tropical Plants, London, 1992

Maxwell, Robyn
Textiles of Southeast Asia, Tradition, Trade and Transformation, Melbourne, 1990

McLeod, M. D.
The Asante, London, 1981

Mason, Philip
The Men Who Ruled India, London, 1985

Miller, Carey D. *et al*
Fruits of Hawaii, Description, Nutritive Value, and Use, University of Hawaii Agricultural Experiment Station Bulletin No. 77, Honolulu, 1939

Montinola, Lourdes R,
Piña, Manila 1991

Moorhouse, Geoffrey
Calcutta, London, 1971

Murphy, Veronica and Crill, Rosemary
Tie-dyed Textiles of India, London, 1991

Payne, Selma and W. J. A.
Cooking with Exotic Fruit, London, 1979

Perenyi, Eleanor
Green Thoughts, A Writer in the Garden, New York, 1983

Piper, Jacqueline M.
Fruits of South-East Asia, Facts and Folklore, Singapore, 1989

Slater, Mary
Caribbean Cooking, London, 1988

Spear, Percival
The Nabobs, London, 1963

Varadarajan, Lotika
South Indian Traditions of Kalamkari, Ahmedabad, 1982

Wilkinson, Theon
Two Monsoons, London, 1976

Zaman, Niaz
The Art of Kantha Embroidery, Dhaka, 1993

What has fruit that bears leaves and leaves that bear fruit? The answer is a pineapple. Philippino riddle

Thanks to...

Geoffrey, Jamie, Amanda and Emily Hyles; Department of Primary Industries, Queensland; Margaret Roberts, Philippa Robertson, Ann Proctor, Natalia Kerbo, Khongthong Nanthavongdouangsy, Brenda Colgan, Annie & Richard Miller, Rosanna Hindmarsh, Sylvia Fraser-Lu, Rhondda, Geoff & Georgie Cleary, Vanessa Griffith, Gek Chong, Jean Lane & Sunset Books, Rosemary Brissenden, Michael Abbott, Carolyn Forster, Diane Firth, Kris Han and the Tea Chapter, Trudy Grabner, Brian Megilley and Del Monte Kenya Limited, Adrienne Johns, Sally Crafter, Patti Mulcare and the Lanyon garden, Maggie Shepherd, Margaret and Neil Cormick, Gudrun Genée, Christine Evans, Victoria Walker, Atty Fe B.Barin and Frances Arespacochaga from Bangko Sentral ng Pilipinas, Patis Tesoro, Anna India dela Cruz Legaspi, Golden Circle and Lindsay Mackerras.

Photographic Acknowledgements

Page 4 Ben Wrigley; page 11 Claudia Hyles, Kate Spring; page 17 Roger Shelley, The Very Nice Group Pty Limited; page 20 Ben Wrigley; page 24 through the kindness of Golden Circle Limited; page 27 through the courtesy of Mr and Mrs Victor Evans, caretakers of the Pineapple House; page 33 Ben Wrigley, Claudia Hyles; page 37 Claudia Hyles; page 41 Geoff Cleary, Sally Crafter; page 42 through the courtesy of the National Museum of Singapore; page 44 Claudia Hyles, Claudia Hyles, Michael Abbott; page 46 Sally Crafter, Sally Crafter; page 47 Claudia Hyles; page 48 Claudia Hyles, Claudia Hyles, Claudia Hyles; page 50 Ben Wrigley; page 51 Claudia Hyles; page 57 Claudia Hyles, Ben Wrigley; page 61 Claudia Hyles; page 69 Claudia Hyles; page 71 Ben Wrigley, Ben Wrigley, Ben Wrigley; page 77 Ben Wrigley; page 78 Ben Wrigley; page 79 Ben Wrigley; page 87 Ben Wrigley; page 90 Ben Wrigley; page 95 Claudia Hyles; page 97 Ben Wrigley; page 99 through the kindness of the Bangko Sentral ng Pilipinas; page 100 Claudia Hyles; page 102 Claudia Hyles; page 103 Claudia Hyles; page 104 Claudia Hyles, Claudia Hyles; page 106 Ben Wrigley; page 109 Ben Wrigley, Claudia Hyles; page 110 Claudia Hyles; page 111 Claudia Hyles; page 112 Claudia Hyles, Claudia Hyles, Ben Wrigley; page 115 Ben Wrigley; page 118 Ben Wrigley, Claudia Hyles; page 119 Claudia Hyles; page 120 Claudia Hyles, Claudia Hyles; page 121 Claudia Hyles, Ben Wrigley; page 122 Ben Wrigley, Claudia Hyles; page 124 Ben Wrigley.